NO EXCUSES, ALWAYS READY

Written and edited by: Robert J. Foucha
Factual accuracy reviewed by: Brian Korte
Book layout and Cover design: IronHull Publishing
Published by: IronHull Publishing

ISBN: 979-8-9935555-0-8
Library of Congress Control Number: 2025917871

First Edition

www.NoExcusesAlwaysReady.com

This journey would not have been possible without the steadfast love and support of my wife, Amber. She has been my anchor through every challenge, and her presence is a source of immeasurable strength. To my children Avah, Micah, and Mason - thank you for your resilience and for bearing the brunt of my service. I recognize the sacrifices you've made, and I carry your love and support with me always.

Table of Contents

Introduction

This book is for the modern Coast Guard professional to give you the tools that you need to become successful in today's Coast Guard. You're coming into a service that, while small in size, is immense in its responsibility and impact. The views and ideas in this book are my own, forged by my experiences during two decades of service in the Coast Guard. My journey took me from a small boat station in the Pacific Northwest to a Maritime Safety and Security Team in New Orleans, to a Dive Locker in Virginia, and beyond. If you are already serving, your views and experiences may obviously differ from my own. I get it. Every duty station, every command, and every mission is different. However, if you can learn one thing from this book, I will have considered it to be worthwhile and worth my time. My goal isn't to tell you exactly how to do your job - your leadership and training should do that. My goal is to give you the foundational wisdom you'll need to navigate your career and get the most out of your time in this service.

The Coast Guard often gets neglected from history and from recollection when speaking about the military in general, however, you and I both know the Coast Guard plays a vital role in today's armed forces. We are a small but mighty force that consistently punches above its weight class. We hold an immense amount of jurisdiction that is vital to enforcing federal laws of the United States. In fact, the Coast Guard has the widest ranging jurisdiction of any federal service, which is something that many other federal agencies and DOD services leverage to accomplish their missions. The authorities we are granted give us a unique position among the armed forces. As written in the Indiana Law Journal, US law grants the "United States Coast Guard and customs officers with virtually unlimited authority to stop, board, and search vessels on the high seas and within the customs waters without any particularized suspicion of wrongdoing, much less a warrant. As such, these provisions comprise what has been accurately characterized as "one of the most sweeping grants of police authority ever to be written into U.S. law."[1] This isn't a theoretical power - it's a reality that we live with every day on the water. It is a great responsibility for such a small service and something that is not to be taken lightly. It's the reason our boarding officers and small boat coxswains must be so well-trained and professional. They are literally the front line of U.S. law enforcement on the water and the image of our country to countless foreign mariners. I take great pride in the fact that, in the Coast Guard, we are not just a number - every person in the Coast Guard has the ability to make a difference. Within your first 4 years, you can be enforcing those laws and making decisions that actually have impact. I did, and I have seen countless young men and women in their first few years

[1] Robert C. Mills, "Note, Border Searches of Vessels: Judicial Confusion of the High Seas with the Customs Waters," Indiana Law Journal 55, no. 1 (Fall 1979): 102.

make decisions that directly impacted lives, property, and national security. There isn't a massively over-inflated workforce, who can afford to just sit around and take a paycheck. Every person counts. The service is too small and our missions too great for anything less. Junior officers and junior enlisted members of the Coast Guard can hold great responsibility when compared to other jobs and/or positions and other military services. This is a blessing and a curse. It means you'll be tested, challenged, and pushed to your limits. But it also means you'll be doing meaningful work from day one, not just waiting for someone else to make a decision or give you permission. This is the reality of our operational tempo and our small size, and it's why our leadership development and training must be taken so seriously.

In today's maritime climate the Coast Guard of the United States and the Coast Guards of the world play a pivotal role. As global trade, resource competition, and geopolitical tensions rise, the maritime domain has become a critical area of interest. A nation's gray-hulled navy is primarily for one thing: a warfighting fleet for naval dominance. Their actions are often seen as escalatory, and their presence can raise tensions immediately. The Coast Guard of a nation is a multi-faceted, multi-domain service with a broad range of capabilities that become the go-to fleet in the maritime realm. From search and rescue to boarding non-compliant vessels on the high seas, from oil spill response to inspecting commercial vessels, from ice-breaking in the arctic and antarctic regions to handing out citations for not enough life jackets on a small boat - the missions are wide and plentiful. We are the workhorse of the sea. As seen in the South China seas, nations' Coast Guards are being put to the test as a means of assessing a country's capabilities and political motivations. This is a critical development in modern geopolitical strategy. In Taiwan, their first line of defense against China is their Coast Guard - and this will play out more all over the world. We are seeing more and more of these “gray zone” tactics used around the globe. If a nation's warship encroaches into another country's waters this will be an act of war - whereas when they use their Coast Guard vessel with wide jurisdictional capabilities, this can be construed as law enforcement or wildlife protection activities. It's a way to probe another country's resolve without crossing the threshold of armed conflict. I truly believe the US Coast Guard and its partners around the world will be more active in the world wide theater in the next 20 years than ever before, due to the increased capabilities of our assets and our ability to provide joint partnership on a global level. The tensions in the Indo-Pacific theater will only increase, and our presence and assistance with our allies will be desperately needed in the coming decades. This means you will be on the front lines of this global strategy. Whether you're on a cutter in the Pacific or a small boat in Florida, your actions contribute to our strategic posture and our ability to project influence. Your professional demeanor and competence directly impact how our allies and adversaries view the United States.

I have been fortunate to be stationed at a DOD command and I can tell you firsthand that the Coast Guard is an amazing organization when compared to all other services. I spent three years working side-by-side with members from every branch of the military, and what I saw solidified my belief that we are, pound-for-pound, the most capable and professional service in the armed forces. What the Coast Guard doesn't have is a massive marketing budget to plaster our image on TV or billboards. We don't run flashy, big-budget ads. I've always had heartache about this and the reality is it's a money issue. The Coast Guard's budget pales in comparison to DOD budgets, so they get Hollywood-style ads while the Coast Guard does what it can. Our recruiting budget is a fraction of what they spend, and it shows. Marketing is a powerful tool that influences people's perception more than they realize. There are entire books, college degrees, and careers dedicated to swaying people's opinions with marketing, and the fact is the Coast Guard typically does a poor job at it. With that being said I'm personally okay with being a silent professional and doing a job that I know may not get great accolades by the public or by the other services, and I sleep well at night knowing our missions contribute directly to the United States and its people. My first tour was at a small boat station, and there wasn't a single day that went by that I didn't feel like I was making a tangible difference in my local community and the nation as a whole. It's a very different feeling from being part of a larger, less-connected force. If you ever have the experience of contributing in a SAR case and seeing a father's face when you rescue him and his son from the middle of the ocean in their small boat, where if you weren't there, they risked capsizing and being lost at sea - you will know what I mean. I've been there. I've seen that father's face and felt that raw sense of relief and gratitude. It's a feeling that stays with you forever and defines why we do what we do. Directly serving the American people is a privilege few military services can claim, and it's what sets the Coast Guard apart. That legacy is worth protecting.

The Coast Guard is not just about getting through Boot Camp. Boot Camp is a foundation - it builds the core skills, mindset, and discipline every Coast Guardsman needs. But the real test comes after, when you step onto a cutter, a small boat station, aviation unit, or any other command where the service needs you. That's where your credibility, qualifications, and standards will either make or break you. This book will walk you through every stage: preparing for Boot Camp, earning qualifications, building your reputation, taking on leadership roles, and navigating the long game of advancement and career choices. I will also show you how to think beyond the uniform - because whether you do four years or twenty, what you build here will define the rest of your life.

Lastly, this book is not intended to be a memoir. It's a guide. Everything in here is written to give you blunt, usable lessons that will actually help you succeed in the Coast Guard. The goal isn't to impress you with stories, it's to prepare you with tools. I've lived the life you're starting, and

I want to pass on what I learned the hard way.

Each chapter focuses on one core area you'll face - from surviving Boot Camp to earning credibility, from leading your first team to planning your long game. By the end, you'll have a playbook you can rely on when the pressure hits and the decisions count.

You'll see four elements at work in each chapter:

BLUF: Bottom Line Up Front. A one-sentence opener that tells you exactly where the chapter is headed - what to expect, and why it matters.

Reality Check: blunt, mid-chapter callouts that cut through the noise. No sugar-coating, no feel good fluff. The "Fouled Anchor" graphic above these sections represent lessons learned from experience - both good and bad.

After Action: the chapter's debrief. It condenses the main points into a clear summary.

Courses of Action: the orders you walk away with - specific, immediate, and non-negotiable.

Throughout the book you'll also find checklists, side-by-side examples, and case studies. These aren't filler - they're tools you can copy, memorize, and use. They're designed to be a quick reference when you're tired, stressed, and don't have time to guess.

The Coast Guard will challenge you physically, mentally, and professionally. This book is meant to be a field manual for success - something you can go back to when you need a reminder of what matters and what works.

Wherever you are in your journey, I wish you the best. I hope for your success in this great organization. I hope you find the same fulfillment and pride in the work that I did. Semper Paratus!

After Action: The Coast Guard is small, underfunded, and overlooked, but its authority and responsibility are massive. From the day you step on deck, you're held to a higher standard because there's no room for excess manpower or wasted effort. You will be relied on sooner than you think, and your performance directly shapes how the public, allies, and adversaries view the United States. Recognition is rare, accountability is constant. You signed up to carry weight that matters - embrace it.

Courses of Action:

- Accept that nobody is coming to hype you up - do the job anyway, with no excuses.

- Carry yourself squared away 24/7 - you're representing the service every time you're seen in uniform or out.
- Take training, standards, and qualifications seriously - you'll be trusted before you feel ready.
- Never forget the Coast Guard's unique role: you're law enforcement, lifesaver, and defender all in one - act like it.

Becoming Always Ready

BLUF: Readiness isn't a slogan - it's a daily standard that separates professionals from pretenders.

Preparing for Basic Training

There are 2 types of people: trained and untrained. The Coast Guard isn't going to spend eight weeks trying to physically get you into shape. They're going to break you down mentally and physically, and they expect you to show up with a solid foundation. If you show up fat and out of shape, you're not just hurting yourself - you're putting a target on your back and on your shipmates. You have to make the conscious decision to become one of the trained.

In every sport, every career, every hobby, and every subject - there was a day 1. That's a day when you were an outsider, not a part of the team, and you had to earn your place. You show up not knowing the ins and outs, not knowing your full potential, not knowing how far you're willing to go, not knowing how much you really care, and the list goes on. The unknowns are massive. This is where the Coast Guard's Core Values come into play: Honor, Respect, and Devotion to Duty. Honor means you're honest with yourself about your training, Respect means you show up ready for yourself and your shipmates, and Devotion to Duty means you're committed to the mission and the service from day one. But you have to show up to figure these things out.

The proverb goes, "the journey of a thousand miles starts with a single step." It was true then and it holds true now. That first step is the decision to train, to prepare, and to dedicate yourself to being ready. Some days you may not want to show up, and that is ok. That feeling is normal; it's a sign that you're pushing against your comfort zone, which is exactly where growth happens. What do you do then? Show up anyway. Even the great toughman David Goggins admits that some days before a run, he doesn't want to go out. What does he do? Sits there, waiting to put his shoes on, sometimes up to 30 minutes. But eventually he puts them on, and gets it done. That's the attitude you have to have when you embark on a journey like this. It will not be easy all the time. It may very well be the hardest thing you've done in your life. You might love it and want more, you might hate it. But you will never know until you do it.

You can take the decision to train lightly, skip those workouts, continue to party with friends before you go, and eat like crap before you show up, maybe thinking that Coast Guard basic training aka "Boot Camp" is a walk in the park and a joke - unfortunately, you will be sadly mistaken. Here's the harsh reality - you will be tested in one way or another. You will be tested physically by the rigorous demands of daily physical training (PT) and the constant "incentive training" (IT) you will get for every mistake your company makes. You'll be tested mentally by the relentless pressure, the constant noise, and the feeling that you are never doing anything right.

In addition to being personally tested, you will begin to have an understanding of what it's like to be on a team and have other people count on you. This isn't just about you anymore. When you fail, the team suffers. The Company Commanders (CC's) will punish the entire company for the mistakes of one individual, which means your unpreparedness directly impacts everyone around you. If you want to ensure your place at the bottom of the team and have people NOT count on you - then don't train one bit. If you show up out of shape, you're not just letting yourself down; you're actively contributing to the misery of your shipmates.

You can choose to be the untrained. You can suffer your way through every evolution, skating by, doing the bare minimum, and getting gassed during every session. You can make this choice, and believe me, many do. They may not make it consciously, but by failing to train, they are making that decision. The sad part is that it's obvious to everyone around except them. People that fail to train often blame other people or circumstances for their shortcomings. They blame the staff, the weather, the gravel on the ground, their shoes, sickness, lack of sleep, not enough water, too much water - the list goes on.

The hard truth is it's not easy to see when you've been lying to yourself or been getting coddled your whole life. Sometimes the people in our lives love us so much they can't tell us the hard truths for fear of reprisal, so they let us keep our ways, even at the detriment to ourselves. This is where you have to look in the mirror and have a hard talk with yourself. Am I ready or not? If you've never hit exhaustion - trembling, muscles failing, and still needing to find strength to continue - you're probably not ready. If you've never pushed yourself outside of your comfort zone to the point where everything in your mind and body is telling you to quit but you still keep going - you're probably not ready.

Preparation isn't about bragging rights in the gym - it's about whether you'll keep your head when you're three weeks in, exhausted, and the whole company is paying for your mistakes. You can't cram for this. A 60-day prep plan before shipping out is the bare minimum: daily pushups, sit-ups, pull-ups, running, and swimming. If you can't pass the run or swim now, you won't suddenly find some newfound endurance in Cape May.

Most recruits underestimate the mental side. Boot Camp strips away comfort, sleep, and routine. Stress control is a weapon: breathing drills, visualization, even just learning to function on broken sleep. Treat those as training tools, not soft skills.

And don't ignore family prep. If you leave behind unrealistic expectations, they'll bleed into your focus. Make sure your people know what's coming. You will have almost zero contact with home so get your affairs in order. When you get that rare chance to call, the roughness in your voice will be obvious, but so will the growth that makes you almost unrecognizable.

Your whole life up to this point has almost always been about you. However, this journey you're embarking on is no longer a solo sport - team is everything. Your shipmates are everything. Your unit is everything. Serving the people of the United States of America is everything. You owe your shipmates, your team, your unit, and your country to be ready for this challenge. The Coast Guard motto, Semper Paratus (Always Ready), is not just a slogan - it's a sacred promise. How can you be Semper Paratus if you're not trained? No more excuses. Time to pay what you owe. You now represent a long, revered line of maritime professionals - both on the sea and in the air - that have ALWAYS exceeded expectations and arose to the challenge. When people need us, we are there. When others can't go out, we do. This is a tradition that spans over 235 years, from the Revenue Cutter Service to the modern Coast Guard - the nation's oldest continuous seagoing service, a legacy we take immense pride in carrying forward. Do not take this lightly. Hold yourself to the highest standard - the people who see you in blue, and the people who depend on you, already expect it.

Reality Check: The public doesn't know the difference or care what rate you are, what rank you are or how long you've been in, they only see one thing - you are a member of the Coast Guard. You wear blue and represent all of us. When they see you, they expect professionalism and expertise. This is a heavy burden, but it's one you signed up for. Wearing this uniform means being ready for more than your assigned job - it means being trusted to save lives, safeguard the seas, and stand the watch whenever the nation calls.

Physical Fitness Prep

Coast Guard Boot Camp is located in Cape May, a small quaint town along the coast in New Jersey. It's hot in the summer and snows in the winter. Aside from being the location of Coast Guard Training Center Cape May, it's a great vacation spot due to the beaches, boardwalk, and amazing homes along the waterfront.

Boot Camp is a required rite of passage. On a practical level, it's where you get issued uniforms, learn the rates and ranks, learn basic seamanship, and transform yourself from a civilian to a member of the World's Greatest Coast Guard. On the other hand, it's where you learn about the history and heritage of the Coast Guard, learn about the legacy and traditions, hear the amazing stories of the heroic rescues of those that have served before you, and ultimately develop a deep sense of pride while learning about the deep history of the service.

What it's not is a fitness camp where you go expecting to have some kind of fitness transformation. It's 8 weeks long. There's only so much you can do in 8 weeks, and building the body of your dreams ain't it. Your goal should be to show up in shape, ready to take on whatever beatings come. And they will come. You will be held accountable for every mistake, every missed detail, and every moment of weakness. Some people show up thinking, 'It's just the Coast Guard, how hard could it be?' The answer depends on how much you trained before you arrived. If you fail to train prior to showing up to Cape May, it will suck and it will probably be the hardest thing you have ever done in your life. If you train, you might actually enjoy it (a little).

I trained for Boot Camp and I can recall knocking out pushups (which you VERY often do) feeling strong, and looking over at my shipmates struggling - and struggling badly! Arms shaking, body damn-near convulsing, making sounds out loud (ehhh, ohhh, ughh - please don't do this - nobody wants to hear you or cares), while meanwhile I'm knocking 'em out. Obviously, everyone has a limit and I hit mine often as the pushup position is a common one at Boot Camp, but had I not trained up prior to - it would have been a miserable experience, as most people seem to have. My limit was in the 75 range and if you want to feel strong you should aim for a high pushup count.

What I'm about to write is not all inclusive. What worked for me may not work for you, so tailor as needed to suit your body and style. I'm not a personal trainer, but I've survived some of the Coast Guard's toughest pipelines - Navy Dive School (NDSTC) three times, DSF Assessment and Selection (discontinued), and more beat-downs than I can count that were meant to break people. What I share comes from experience, not theory. Later in my career, I served as a high-risk training instructor at Navy Dive School. I ran evolutions that pushed candidates to their limits and saw grown men and women quit on the spot - choosing DOR (drop on request) rather than endure the demands. I've been at the breaking point and kept going when there was nothing left to give. I've pushed people to their breaking point. The difference between quitting and continuing is mental fortitude - nothing else.

Some quit when the pain is too much. Some suffer and continue on. If you fail to train, it will suck 100x worse and it will be painfully obvious to yourself and everyone around you. People will remember how badly you suffered and will lose confidence in you. Deep down they know you weren't committed to the team because you CHOSE to not prepare. You chose the easy route and when things get hard you will choose that path again. That speaks volumes of who you actually are and your work ethic. If not consciously, people see this subconsciously and judgments are ultimately made. Here are the basics to train you up.

HYDRATION

Hydration is one of the most overlooked factors at Boot Camp, yet it's one of the easiest things you can control. When you show up to Cape May, you're packed into a squadbay with recruits from every corner of the country. Different states, different environments, different germs. Within the first few weeks, just about everyone will catch what's known as the "Cape May crud" - the cold-like sickness that spreads like wildfire when you stick 100 people in close quarters running on stress and exhaustion. You can't avoid it completely, but you can stack the odds in your favor, and one of the strongest defenses you have is staying properly hydrated.

Most recruits underestimate how much water they actually need. You're sweating through PT, running in the humid New Jersey summer or freezing in the winter, marching, drilling, and living on your feet. Dehydration doesn't just make you tired - it weakens your immune system, dries out your mucous membranes (the first line of defense against germs), and slows down recovery after hard training. A dehydrated body is an easy target for infection, and once you're sick in Boot Camp you'll fall behind fast.

Hydration also directly affects performance. Even mild dehydration - as little as 2% of your body weight in water loss - can cut your physical performance and mental sharpness. That means slower run times, sloppy drill, poor focus in class, and more mistakes when every detail is being judged. You don't get second chances in Boot Camp. Falling out of a run or zoning out in a lesson because you're dried out will cost you credibility you can't afford to lose.

The solution is simple: make water a priority. Drink at every meal, fill your canteen and actually use it, and don't rely on soda or caffeine to "hydrate." Electrolytes have their place, but nine times out of ten plain water is what you need. If you treat hydration as a discipline instead of an afterthought, you'll recover faster, fight off illness more effectively, and perform like someone who belongs there - not someone making excuses for why they can't keep up.

Pushups

You should be able to do 50-70 push-ups on demand. If you can't do this, you will struggle. The pushup is the classic military go-to for all things physical. Don't do something right? Pushups. Warming up for a physical activity? Pushups. Trying to make you stronger? Pushups. Trying to break you down? Pushups. Classic. It's the most common physical punishment and exercise in Boot Camp, and you'll do them constantly - with the entire company and when you get targeted directly.

Before attending basic training I set a timer to go off every hour and I would do 20 pushups, no more. Can't do 20 starting? Do 10 pushups every hour on the hour, no exceptions. Set an alarm on your phone and stick to the plan. Do this for a minimum of 90 days prior to you showing up. By the end of this, if you actually do this, you will be stronger and should be able to knock out 50 in one setting with no issues. See **Appendix A** in the back of the book for a complete routine.

But what about females? Isn't the standard different? Look. The numbers on paper don't tell the whole story. I've seen women across the Coast Guard and Navy, from high-risk specialties to conventional roles, flat-out outperform the men. Some went on to become Coast Guard Athletes of the Year. Others weren't looking for recognition - they just showed up every day and crushed their job. While specific physical benchmarks may vary, the **universal standard of fortitude** is what defines these professionals. Gender aside - determination, training, and getting after it is all-inclusive.

As fate would have it, I was later stationed with one of my Company Commanders, QMC→BOSN→LT→LCDR (select) Greg Isbell - the baddest, meanest son-of-a-bitch that I had ever met. He set the tone for my entire career to come. He was Semper Paratus to the core and it was everything I needed. If the Coast Guard had a fleet of Greg Isbells - the whole damn ocean would've been at parade rest! I served as a squad leader for a time, which meant taking extra IT (incentive training) when the company fell short. Even after being relieved of the role - a common occurrence - I still stepped forward when the call came: "SQUAD LEADERS, DOWNSTAIRS, NOW!" I took the punishment whether I wore the title or not. I was here for ALL of it! Greg Isbell, once I later knew him personally and not only as a menacing, smokey-the-bear brim wearing punisher - turned out to be one of the most awesome and most squared away Coast Guardsman that has ever worn the uniform. He commanded respect the second he entered the room and he was THE standard. The man was the <u>definition</u> of squared away, with a career that included years more years as a Cutterman than most will ever get, BOSN of an MSST, Command Duty Officer for District 17, XO of a cutter, and CO of a patrol boat (and much more). Don't show up weak - or LT Isbell will come out of retirement and be waiting in Cape May to IT you back to the stone age! No Excuses.

Reality Check: Pushups build upper body strength. You need to be able to save yourself first. If you fall off the ship, fall off the boat, need to climb back into the vessel or traverse a jacob's ladder to board a ship - you NEED to have upper body strength. The ocean is unforgiving. It does not care about your gender, your age, your background or anything else you think makes you a special unicorn - it will swallow you up. Upper body strength is crucial to saving yourself.

CORE STRENGTH

This isn't about having a chiseled 6-pack of abs, this is about having core strength. Your core is the foundation of your body, and without good core strength, your body will be more prone to back injuries and you will struggle with the physical exercises throughout your career. You should be able to do 50-70 sit ups in two minutes and hold a plank for at least 1 min 30 seconds before you show up. Yes, I know this is more than the "recommended" minimum, but I never did the bare minimum, and neither should you. The bare minimum is the lowest bar - and if you settle for that, expect the lowest results in life. If you want more, push yourself. Always aim above and beyond, in every category.

Company Commanders will push you to failure at some point in almost every exercise you do, and if all you're used to is the bare minimum and expect a break - you'll be in for a rude awakening. Again, do not wait till you're there to realize you should have trained. Your core strength is one of the most important indicators of overall muscle health in the human body. Start training your core now and you'll save yourself injuries later. I learned this the hard way - after multiple strained lower back injuries, I realized my weak core was forcing pressure onto my spine. Your core isn't just abs; it wraps around your entire midsection, protects your organs, and stabilizes your spine. Neglect it, and your back will eventually pay the price.

Your goal should be to pace yourself so as to not burn

out too fast, but go fast enough to get the max number of reps in the allotted time. You should develop a fast cadence at first then pick up the pace. 2 minutes is not a lot of time, so you need to knock them out quickly. Do not forget to breathe. If you hold your breath while doing situps, you will get gassed quickly - breathe.

Flutter Kicks

This exercise is when you lay down on the deck and lift your legs in an opposite sequence like you're swimming. Flutter kicks are often performed with your hands palm down under your buttocks to support your lower back. You begin by lying down on your back with your head off of the deck, looking forward. You will likely have on tennis shoes ("go fasters") the first time you perform this exercise, but it will inevitably lead to doing this in full uniform with boots on. Yes, steel toed boots are significantly heavier than tennis shoes, so don't be ill-prepared when it comes to this. This exercise will crush your hip flexors, so be prepared for this simple, yet strenuous exercise.

Flutter kicks are often counted by the 4 count (1,2,3 - 1. 1,2,3 - 2). You should be able to work up to 50 flutter kicks in a row. Another brutal part of this exercise always seems to be at the end, when the call for "6 inches" is made - meaning, keep your feet straight out in front of you, 6 inches off of the deck. How long do you keep them there? Until you are told to stop. These little rituals are meant to strengthen you, punish you, toughen you up, and increase your tolerance to pain. After performing flutter kicks, and when you're smoked, and that call comes for "6 INCHES!", you will quickly know who actually trained for this, and who spent all of their free time hanging out or playing video games.

Pull Ups

You may not do pull ups all the time, but when it's time to mount the bar and do a dead hang pull up, that's not the time to realize you should've trained on this. Most people struggle with pullups because it is one of the ultimate measures of upper body strength. You WILL require upper body strength in your career, so do not neglect this. It may be the difference between you saving yourself, your shipmate, or someone else.

Imagine if you fell overboard and they lower down the net for you to climb back up, but you can't even lift yourself up. Or imagine you're part of a boarding team and you've been tasked with hauling up the boarding kit after you've boarded - but you can't. I can think of a 1,000 scenarios where you need the ability to pull your own bodyweight or bad things can happen. Do not assume you can, train so you know you will. Strive for at least 6 pullups at a bare minimum. Can't do 1 pullup? Start at the gym with the pullup machine, where you can add weights to offset your body weight during a pullup. If you can't already do 1-6 pull-ups this will be hard in the beginning and will take time so don't wait until the last minute. See **Appendix A** for a program. Suck it up and train.

Lats

Use the lat pulldown machine to build up your upper back muscles to get better at pullups. This machine will increase your upper body strength and help you with pullups. A good tip someone once told me about lat pulldowns was this: "Squeeze your lats together, and when you do, imagine you're trying to hold a pencil in between them." I always think of this when doing lat pulldowns. Increase the strength in your lats, increase your upper body strength. This will also help with pushups and overall upper body strength.

Another upper body workout is "chest-to-bar". On a smith machine or horizontal bar a few feet off the ground, place the bar in a position where you can grab the bar with your legs out, grabbing the bar from underneath, resting on your heels and then pull your chest up to the bar. This has been said to be one of the best upper body workouts you can do. You can look this up and find plenty of resources on this workout.

Swimming

Listen. What I'm about to say is a sad reality that I can't

believe even needs to be said - but I'll say it anyway because I've seen it way too many times. You are joining a sea-going service. I'll say it again - **SEA...GOING....SERVICE**. Learn to freaking swim for crying out loud! You'd be surprised how many people in the Coast Guard - and even the Navy - can't actually swim. It's embarrassing. That's like being in the Air Force but afraid to get in a plane, or a Marine who freezes at the sound of gunfire. It's like a firefighter who's afraid of fire, or a surgeon who faints at the sight of blood. You may not be on the water every day, but sooner or later, the ocean will be part of your job - and when that moment comes, you don't get a second chance to figure it out.

If you have a fear of water, you may want to find another way to serve your country - it's kind of what we do. Not all the time for all jobs, but it's always a possibility. No, you may not find yourself in the water often (unless that's your job - Hooyah Deep Sea) but if you do find yourself in the water and you freak out, you might not only kill yourself but your shipmate too. You need to be able to save yourself first and foremost before you can save others - like the reason the airplane announcement says to put your oxygen on first, then help others - you can't help others if you're incapacitated.

Another thing about swimming that I notice is the average person grossly over-estimates their in-water confidence and abilities. I have given countless PT tests for aspiring Divers and have watched many seemingly fit people just stop swimming and get out of the pool. I've also seen highly experienced people who have been in for 15 years get so gassed from the pool they needed an ambulance and oxygen (not a joke). For some reason, people remember the last time they swam, even if it was years ago, and imagine they have the same abilities as they had then. Never mind the fact that it was years ago and you haven't worked out in a year and you gained 30 lbs since then.

I have seen this time and time again, only to see the look of sheer panic and defeat on their face when they are in the pool for the first time in a few years and trying to swim laps or perform some waterborne task under pressure. They get gassed quickly and THEY actually need rescuing in a flat, calm, heated pool. How do you adapt yourself to getting used to a stressful situation - do it over and over again, and do it often! But people always say, "But we wear a life jacket every time we get underway!" To those people - I say you are naive, inexperienced, and are over confident in that cheaply made vest you depend your life on. I've seen countless people WITH life jackets struggle to swim 500 yds in a pool with simulated waves, wind, and rain. They wind up on their backs like a turtle - gassed and aimlessly backstroking without looking where they're going, running into other people, and a complete hazard to themselves and others. Quite the opposite of a maritime professional who is expected to save lives.

I've also witnessed people slip entirely out of their life jackets while struggling in the water, and unfortunately, I've seen firsthand the fatal consequences when CO2-inflated life jackets fail to deploy. More than once in recent history, people have lost their lives while wearing heavy gear during boardings without proper buoyancy correction. These are the policies that are written in blood - our brothers' and sisters' blood. The gear that you wear can only do so much to save you, the preparation and rehearsed use of it during duress is solely up to you. The only definitive way to prepare for water survival is to train for water survival, which starts with survival swimming and getting used to water in your face, up your nose, and in your eyes. That is the base and do not neglect it for one second.

Reality Check: I will be a broken record here - the ocean is unforgiving. Do not take it for granted. Do not assume you are better than you are in the water, you HAVE to train for this. Your life and your shipmate's lives depend on this. Boats can sink, helicopters that fly over water can fall from the sky, and people who join a maritime service can find themselves in the water - do not assume this cannot happen to you. Do whatever you need to do to train for this and do not let this perishable skill lapse during your career. I give you full permission to shame those around you who suck at swimming if they don't train - they are putting everyone around them in danger.

Running

Let's face it - if you weren't already a runner, running sucks. However, it's part of the job so you might as well get used to it and train. Before I joined, I never ran for fun and definitely not for competition. Running wasn't part of my life or anyone around me. I only ran a few times before Boot Camp, just enough to fool myself into thinking I was ready. Truth is, I did the bare minimum. I overestimated my ability, underestimated the standard, and paid for it. Learn from my mistake - don't show up like I did.

I still remember barely dragging myself around the track in Cape May, sucking wind on a simple mile-and-a-half run. Pathetic. Years later, I was faster than ever - because I finally stopped pitying myself when it hurt and learned to push past my comfort zone. I never played sports or had a coach to push me, so I simply never pushed myself. That's not an excuse, it's just the truth. I didn't find my real limits until later in life - and then I was pissed I hadn't discovered them sooner!

If you search the Internet or books you will find more information on running than you could ever process so I will keep it simple and tell you what has worked for me and what I have seen work for others. At this point I've come to realize you are either gifted in running or you have to work at it, and we both know who we are. I have gone from barely passing the mile and a half as a 20 something-year-old to numerous grueling military schools/training to now running mountain marathons in my mid 40s. You can get better if you train and are dedicated to learning how to actually "enjoy" running.

On the days I do not feel like running but I know I have to, I kindly remind myself that my running days are limited (I nicely tell myself "get your MF'in a** out there and get to work you piece of sh**t, you won't be able to run one day") and I remind myself there are people who wish they had two good legs so they could go on a run. That may sound morbid and harsh but it is the absolute truth and sometimes we need to remind ourselves how good we actually have it. I also remind myself that there are some people with no legs running on prosthetic limbs and that I have zero excuses except to put 1 foot in front of the other.

If running isn't your strength, think of training in two categories: speed and distance. When I say distance I mean many miles at a time such as a half marathon or greater. When I say speed I mean for time, as in you have a threshold that you have to meet in order to pass a test. I'll be talking about speed training because that's what we need. For starters, if you have never ran 1.5 miles, go out and do it as a baseline. It's ok to listen to music when you're training but just know that it will not be allowed during your tests. You need to get used to the sound of you pounding the pavement and your breathing. Once you have completed your baseline 1.5 miles you will know where you need to improve.

Interval training has been my personal key to improving my run times while training for my fitness tests. Interval training has been proven to be more effective than distance training for runners (there are scientific sources but I'm not citing them because I'm not trying to win scholarly accolades here). Interval training like Tabata is great for improving performance on 1.5-mile timed runs. By alternating short bursts of high-intensity sprints with brief recovery periods, it builds both aerobic capacity and anaerobic power. This trains your body to recover faster, push harder, and sustain higher speeds - exactly what you need to shave time off a short-distance run under pressure.

My personal method of run training is given in Appendix A in the back. Disclaimer - I'm not a running coach and you may need to find what works best for you.

Another helpful tool when training to run is a watch with a pace calculator such as a Garmin GPS watch or some other watch that can tell you if you are meeting or exceeding a target pace. This is extremely helpful to know how fast you're actually going and if you need to pull back or speed up, based on your target goal. A lot of people will get excited at the beginning of a running test and take off as if it's a foot race and lead the pack for a half to 3/4 of a mile, then get tired on the back half of the run and score horribly because they got tired too quick. You have to pace yourself and finish strong.

Reality Check: Unless you are one of the gifted, running is hard. Suck it up and do it anyway. Cardio endurance is something most of us have to work for. In the diving community there's a joke that "we haven't ran to a dive job yet" as a way to complain about running, but the truth is we all know how much it sucks to be winded during a dive while finning through strong currents and that our cardio endurance is crucial. As I said before, when you wear the uniform the public expects you to be trained. They think of the Coast Guard as a rescue service although we know there's so much more. If you show up somewhere and can't perform, can't assist, or give up because you're too tired to continue, I pray that no one captures that on video because it will be on repeat for years as an embarrassment to yourself and more importantly to our service. Stop being weak, put on your damn shoes and get out there.

Shoulder and Triceps Endurance -The Forgotten Prep

If you've only been training pushups and situps, you're not ready. Boot Camp isn't just about max reps - it's about pain management and endurance in positions designed to break you down mentally. One of the most brutal examples is the overhead hold, along with front and side holds that torch your shoulders. At Cape May, you will spend hours holding canteens, rifles, or whatever else the Company Commanders decide belongs above your head or out in front of you. It's not about the weight - it's about the time. Ten minutes feels like an hour, and an hour feels like hell.

The biggest mistake I've seen, and one I made myself, is underestimating how much shoulder and triceps endurance you need. Pushups won't save you here. Bench press won't save you here. You need direct work for deltoids, traps, and triceps that conditions your muscles to stay locked in under fatigue. That means training overhead presses, lateral raises, triceps dips, and extended isometric holds with light weights or even water bottles. Build your tolerance to the burn before you ever set foot in Cape May.

This isn't about looking good or moving big weight - it's about surviving the endless "hold it up" games that every company goes through. The ones who quit first in these drills aren't necessarily out of shape - they just never trained their shoulders for endurance. Don't be that person. Prepare for it now, and you'll save yourself a world of pain later.

Caffeine and Sleep Discipline

Caffeine hangs around in your body a lot longer than most people realize. The half-life is about five to six hours, which means if you crush an energy drink at 1400, half of it is still in your bloodstream at 2000. But the story doesn't end there - it can take 10 to 12 hours for your body to fully clear it. That means caffeine consumed late in the day is still working against you well past midnight.

I learned this the hard way. There were plenty of nights where I went to bed thinking I'd finally crashed, only to snap wide awake at 0300 - wondering why I'm up. It took me way too long to connect the dots back. That "second wind" in the middle of the night wasn't my body magically being rested - it was the caffeine still circulating, keeping my brain lit up when I needed deep recovery.

Sleep is the most underrated part of being always ready. You can lift, run, and study until you're burned out, but if you're not letting your body recover, you're wasting half the effort. The fix is simple but non-negotiable: set a caffeine cutoff. For most people, that means no caffeine after 1200 if you want a real night's sleep. You'll perform sharper, think clearer under stress, and avoid burning out before your career even gets started.

Pain and Suffering

There will definitely be times that you are pushed to your limit and the physical pain of what you're going through will ring out through your mind, and every fiber of your being wants the pain to stop. That's a normal human response to pain - your mind tries to avoid it at all cost and, yet, here you are - in pain and you are not in control of when the pain stops. There will be times when it sucks, there is no way to sugar coat it.

Suffer in silence. Moaning, groaning, crying, whining,

making noises, grunts, yells, whimpers or any other vocal outbursts do not help. If anything, you are further demoralizing your team. During these moments of intense and unavoidable pain, no one can save you, no one can help you, and no one will feel sorry for you. If anything, by making all these noises and sounds, people will start to question your resolve and whether you have what it takes or not.

I've been in both positions many times. I've been on the receiving end of some brutal physical "beat down" sessions, and it always rubbed me the absolute wrong way when one of my classmates would enter into this "poor me" vocal session when they reached some physical pain threshold, where they would appear to be crying out for help, begging for mercy, and complaining all at the same time. It has since become kind of a pet peeve of mine. As a high-risk training instructor at Naval Diving and Salvage Training Center (NDSTC aka Navy Dive School), I've also been on the giving end of many physical remediation sessions and I've seen people from all of the services play this game of "oww it hurts" when in the thick of it. Suck it up and stop feeling sorry for yourself.

Working at a joint, multi-service training facility that trains people from all services in military diving (except SEALs and Army Special Forces), you see people at every age level, from all walks of life, from all of our nation's services. However, at the core of it, they are all just people. I've been through enough PT-sessions to get a real-world degree in making people suffer and there is always that someone in the group. When you're exhausted from doing more burpees than you thought were humanly possible in one setting, and your arms are shaking from being in the "lean and rest", and you're making pools of sweat on the pavement beneath you as it drips down from your face, and you hit that mental and physical wall that screams in our mind and says "make it stop" - you will have a decision to make - to cry out for help, beg the mean people to make it stop and let everyone know just how weak you really are and how much pain you're in - or - suffer in silence. You should always choose the latter.

Your mind will quit 100% of the time before your body does. You will always get that urge in your mind that says, "Um, I don't like this. Do we have to do this?" It's in our nature to avoid pain - and yet here we are voluntarily putting ourselves in pain. This makes us feel like we have a choice in the matter, when the truth is - we don't. Yes, it will hurt. No, you probably won't die from it (probably not). Yes, you are capable of 20X more than you think you are capable of. When your mind feels the pain from your body you need to go inward and not project outward - the simple truth is: nobody cares. The instructors don't care that it hurts - they know it does, that's why they're doing it. Your shipmates/ classmates don't care - they know it hurts too, they're doing it with you. Don't give them a reason to doubt you by crying out for mercy and begging for sympathy, because that's how it sounds to everyone else.

How long will the physical pain last? You don't know, do

you? That's exactly what they want. The unknown makes the pain worse and messes with your mind more. Sometimes, they want you to quit the voluntary course you're in, just to see if you will. I've also seen that more times than I thought I ever would. You need to remember this - they cannot go on forever. There will be an end to the evolution. When? You won't know and it probably won't be when you want it to - but the end will come. They cannot PT you forever. You just need to focus on one evolution at a time - that's it. Get through that one evolution. When the next evolution starts, focus on getting to the end of that one. One by one. When people start thinking too far ahead, it becomes overwhelming, then you start to doubt yourself and wonder if you can continue this for that long (X more weeks, X more months, etc). When you start feeling overwhelmed, you start to spiral in your mind and some people can't come back from that. One evolution at a time.

When you think you're at your limit and you need an escape - you have to find your happy place. You need to make a mental image, scene, or place in your mind that you can escape to when the suffering gets too intense. You may think it sounds cheesy, but trust me - I've gone to my happy place more times than I'd like to admit, and it got me through beatdowns that felt endless. When it hurts, simply close your eyes, imagine your happy place and suddenly you're transported to another dimension where the pain doesn't matter anymore. Even if it's for just a few seconds of relief, your happy place can help you hold out for just a bit longer.

Everyone's happy place is different, so you need to have a mental exercise to find your own, preferably before you need it. Then when you need to go there, boom - teleportation into another world and some relief from the mean people who want to hurt you. While you may choose to not have your very own happy place and you prefer to live in the moment, that's up to you. I'm just offering tools for your toolbox based on my own personal experience, which may not be your thing, and that's totally fine by me. However, one thing that's not fine by me, is to hear your whiny self when it hurts. Do us all a favor and suffer in silence.

After Action: Nobody said this would be easy. If you want easy, you're choosing the wrong career path. There are plenty of 9-5 jobs out there that will accept mediocre performance and be perfectly ok with you doing the bare minimum - this is not that. You will need to put out and give this your all. Your goal should be to leave it all on the grinder - to be the most HOOYAH person there, as I tried to do. You may never get these chances to put out again - ever. Don't look back and wonder what you could have done - go hard, get some, and get after it! When you start to feel sorry for yourself because it hurts - and it will - suck it up. Nobody cares - work harder. You have to earn this, day in, day out, and learn to push out that overwhelming fear out of your mind when you start to feel it. Focus on the evolution - get to the end. Next evolution starts? Focus on the end of that - get the end. Rinse and repeat. It's that simple. No one will feel sorry for you if it hurts. Smile in the face of adversity and never give them a reason to know it hurts.

Courses of Action:

Boot Camp Prep – 60 Days Out

- **Train before you show up:** 50 pushups, 50 sit-ups, 1.5 mile run in under 11:30, 500-yard swim in under 11 mins. If you can't hit these, you're not ready.

- **Build real endurance:** push past comfort, train shoulders, core, and cardio until failure isn't foreign to you.

- **Control the basics:** hydrate, cut caffeine early, enforce sleep discipline, and prepare your family before you ship out.

- **Toughen your mindset:** learn to suffer in silence, focus on one evolution at a time, and stop looking for sympathy.

- **Knowledge:** Memorize the Eleven General Orders, Core Values, rates/ranks, and other knowledge. These aren't trivia - they're weapons against humiliation on the quarterdeck. Search: CG Recruit Training Pocket Guide

See **Appendix A** for detailed plans.

Earn Your Credibility

BLUF: Credibility comes only from consistent performance, not talk.

Whether you're fresh out of Boot Camp reporting to your first unit or on your twilight tour (your last duty station before retirement), we all have one thing in common - we must be qualified to be a contributing member of the unit. The moment you step onto the quarterdeck, your command is already looking at you as an asset they need to utilize. They've been waiting on you, sometimes for months, to fill a critical gap on a duty rotation or a watch bill. Every unit requires its members to become qualified in their respective duties in order to be a productive and trusted member of the unit. This isn't just about being a good shipmate; it's about operational readiness. If you're unqualified, you're a liability. You're taking up space and resources without contributing to the unit's core mission. It may be becoming a qualified boat crewman at a small boat station, getting Officer of the Deck (OOD) qualified on your cutter, or getting qualified as an aircrew or rescue swimmer at an air station.

Our unofficial motto in the Coast Guard is "do less with more" and that means less people to do more jobs - so everyone is that much more important. This is a cold, hard truth. The Coast Guard is a small service with a massive mission set. We don't have the luxury of carrying extra weight. Every single person on a boat, a cutter, or in the air has to be able to pull their weight and perform their duties when the time comes. This isn't a 9 to 5 job where you can hide in a cubicle. When the alarm sounds, your team is depending on you to be ready to go. The person next to you is literally trusting you with their life. In the Coast Guard, and every organization for that matter, there is always "that guy" (it's just a phrase and you can definitely be "that girl"). "That guy" who doesn't share their equal load of responsibilities - "that guy" who takes way too long to get qualified and they're not put into the watch rotation, making the duty schedule worse for those who actually did the right thing - you get the picture.

Upon arriving at your first unit or a new unit or to a new watch section, your main mission in life should be to immediately get qualified. You need to hit the ground running. You'll likely be assigned a mentor or work with a training petty officer. They are there to help you, but they aren't there to hold your hand. This is your career and your reputation, and you need to drive the process. The faster you get qualified, the faster you will succeed in this organization. You won't just be an extra body - you'll be a trusted member of the crew. Upon arriving at my first duty station it took me way too long to figure that out. I was "smoking and joking" with the other non-rates and was acting like the station was a college party place. The seniors saw it, and they were watching. Every missed opportunity to learn, every time I chose to hang out instead of study - it was noticed. I was a break-in crewmember for a while, which basically meant I had to have adult supervision every time we went out on the boat. This wasn't just a slight inconvenience - it was a huge drain on the unit. Every time we went out, an extra qualified person had to be assigned just to supervise me. Although I was sent to the station as a +1 to add to the crew, I was actually making more work for the crew because they not only had to teach me but had to babysit me until I got qualified.

Once I kicked myself in the rear and realized the name of the game was qualifications, I quickly rose from being a self-initiated dirtbag to a star performer at my unit. I started to study relentlessly, became obsessed with getting qualified and in no time started to make the other new guys look bad in the process. That wasn't my intention but it became obvious to the leaders who was putting in the work and who wasn't. The leadership in the Coast Guard is always looking for the motivated, driven members. They see who is taking the initiative, who is asking for opportunities, and who is putting in the time. They're already sizing you up for future leadership roles and advancement. To put it bluntly - you are almost worthless to your command and to your team members if you are unqualified. You are dead weight. The longer you are not qualified the more your team has to pull your slack while you try to qualify. The longer you do this the more resentful your team will become of you for dragging your feet, and people don't forget these things. This is a small service. People will know your name and they will know your reputation. It will follow you from unit to unit. In a service of small teams - your reputation is everything. Before you get to a new unit people start asking about you - your reputation precedes you - good or bad.

Your first 90 days at a unit will brand you. Show up late, miss watch turnover, or stumble through your quals, and you'll be remembered as unreliable. Fixing that reputation will take years, not weeks. A simple timeline saves careers: Day 1 to 30: learn names, learn the routines and schedules, start PQS. Day 31 to 60: qualify on your first watch station. Day 61 to 90: prove you can be trusted without someone holding your hand.

Credibility bleeds directly into advancement. If you drag your quals you'll miss timelines, evals will suffer, and you'll get left behind. Leaders keep score on effort - they know who's moving and who's standing still. Don't kid yourself - the clock starts ticking the minute you report.

At your unit, you will have a grace period and a timeline that it should normally take for you to get qualified, but you should not take this for granted and you should not take this timeline lightly. The command has set that timeline for a reason. It's a reasonable expectation of when a motivated and competent person should be qualified. If you fail to get qualified in that timeline there may be negative consequences to bear. This isn't just a threat - it's a reality. You could be denied advancement, get a poor evaluation, or even be transferred to a different unit if the command feels you're not cutting it. The fact that there can be negative ramifications for you not getting qualified on time should tell you the severity and importance of you becoming a productive member of this team. Your team will literally have to work harder because you are not qualified. Still a break-in boat coxswain? Your unit has to supply an extra person every time that boat gets underway. Still not qualified as a flight mech? Your unit has to supply an extra mech with every flight that takes off. That means someone who is qualified is missing out on sleep, time with their family, or other training opportunities because they have to be your babysitter. If you drag your feet during this process or let it be known that this is not your priority, your unit and team will definitely not take kindly to this and you will quickly find yourself as the outcast. You don't join the Coast Guard to be an innocent bystander and watch everyone else work. If you want to be a team player, want to be treated well, and want to be successful with your time in - be relentless about getting qualified.

How To Get Qualified

I once asked my fellow deep-sea diver and badass ultra-marathon-running friend Loren Powers, DVCS (Ret.) a dumb question while we were training together: "How do you do it? How do you run for 100 miles?!" His answer was so simple I felt stupid - "One foot in front of the other." We're the ones who overcomplicate things - the answer is always simpler than we make it. Each qualification (i.e. Boat Crewmember, Communications Watchstander, Aircrew, etc.) has a dedicated manual with tasks that you must complete, and knowledge you must know, in order to attain that specific qualification. That booklet/packet is called the PQS (Professional Qualification Standard). This PQS is the official, codified standard for your qualification. It's not a suggestion; it's the rulebook. You can't just "feel" like you're qualified - you have to be able to prove it.

Note: There is also a process/packet called a JQR (Job Qualification Requirements) that is often considered a workstation-specific qualification (i.e. how to operate a specific piece of machinery that not every unit has) or something that doesn't broadly translate to the entire service - for the purposes of this book they are the same.

Typically, people who are already qualified can "sign off" trainees' PQS; these people are called PQS Signers. It's a system where people who are qualified teach people who are not. This is not all-inclusive and there may be qualifications that require special training and special circumstances to get qualified - I'm speaking generally here. For each task in the PQS, there is a reference, telling you what manual/instruction/policy the requisite knowledge on the task comes from. This is the background of that task and is often a deep dive on the why and how. Sometimes it's a rabbit hole of information but follow the trail and take notes. Do not skip this step. The reference material isn't there to make your life harder - it's there to provide the foundational knowledge you need to be an effective operator. PQS tasks often take the shape of 1) things you need to know (knowledge based) and 2) things you need to do (performance based).

For things you need to know - study. It's literally that simple. There's no secret trick. You have to put in the work. Not good at memorizing things? Nobody cares, work harder. This is a matter of life and death, not a school test. If you can't recall a critical piece of information in an emergency, someone could get hurt or killed. Use physical or digital flash cards, have PDF manuals on your iPad or computer and highlight important parts, print out pages from manuals (what we used to do and everyone had stacks of 3-ring binders - don't recommend now) and study. Use whatever method works for you, but the key is to be active in your learning. Don't just passively read the material. Life is busy, your phone wants your attention, your friends want to hang out and all the other excuses you can come up with will be there. You HAVE to dedicate and set aside time to study if you want to truly succeed in this organization. You will not get qualified on your downtime at work. The unit is always busy with a thousand different things, and you'll always be interrupted. You have to take this home with you and make it a priority. Learning how to study is crucial to your success, as you will learn later when you start taking Service-Wide exams where you take a test of your rating knowledge to compete to advance in rank. Start your time off right and learn how to study tasks and references, as this will be the building blocks of a successful time (dare I say career) in the Coast Guard.

For things you need to perform, the best way to learn is to watch someone experienced do it. Do not let anyone cheat you out of this method. Monkey-see, monkey-do. If you've never seen X accomplished and someone expects you to do it, how do you know the intricacies of the thing? You don't and you won't. You might not get it the first time you see it but repetition is the key here. For example and something I remember fondly - if you've never seen someone take out and setup the anchor on a 47' Motor Lifeboat and you were told to do it - you will without a doubt fail to do it in the correct way (which is the safest way) and done that way for a reason. You will get some choice words when you start to become a hazard to yourself and to your crewmembers in 20' seas in the Pacific Ocean. "But I've been on a boat plenty of times - just grab the anchor and chuck it overboard." Wrong. Take out the anchor with the boat thrashing around in the seas and watch the anchor become a missile and fly into the air when the bow goes up a wave and falls down the face of another. Congrats, you just killed yourself or your buddy with an anchor to the face. The proper procedure involves taking out the line while the anchor is still secured, communicating with the coxswain, and carefully deploying it in a controlled manner. Monkey-see, monkey-do is the way and quite often the best way to learn and understand not only how but WHY things are done the way they are.

Once you have learned the knowledge and can recite the critical points of it back to someone who can "sign off" your PQS task, you can get that particular task signed off. The same holds true to a performance task.

The qualification method is typically this:

- Have a need/want for X qualification
- Get PQS/JQR
- Look up references for a task

    a. Take notes

    b. Study notes

    c. Ask questions from people are qualified

- Watch someone complete the task if you've never seen it happen
- Get task "signed off"

    a. Knowledge Task? → Answer questions from someone who is qualified → Get their signature on that task

b. Performance Task? → Complete that task in the presence of someone who is qualified, and based on their subjective assessment of your performance → Get their signature on that task

- Complete all tasks in PQS/JQR

- Oral Board

- Check Ride

- Qualified

- Maintain qual through re-certifications as needed/ Train others!

Note: This process is not all inclusive and your process may be different. An oral board or checkride is not always required so don't shoot me if you go through some process and it's different. I'm not trying to write a novel here, it's more of a guide.

Looking at a qualification book can be an overwhelming thing. Some of the qual books are not small. There may be hundreds of tasks to perform and an overwhelming amount of information that needs to be learned and/or memorized. You have to have a plan or you are bound to get overwhelmed, stressed out, and fall behind schedule. You can't just take it one random page at a time, you'll get lost and fall behind. You have to break it down into chunks and make it digestible. You should be looking at the book and prioritizing the most important tasks first. You also have to take every opportunity that you can to gain experience. I'm not going to go into a tirade or lesson about how humans learn (kinesthetic/hands on, visual, etc.) - the truth is you probably already know what style works for you and that's all that matters at this point. With every job and responsibility that I've had, it almost always took knowledge and memorization of critical items combined with proven hands on ability. You don't have to get it right the first time and you should not expect yourself to or you'll be setting unrealistic expectations for yourself and setting yourself up for failure in your mind.

For every qualification task there is a reference. The reference is the background knowledge you need to know and explains what you need to know to complete the actual task. Go to the reference and highlight or underline key items. In the Navy, people write out short notes directly on their qual page, showing the person who has to sign off your task item that you actually did the research and homework. I've never seen anyone in the CG do this, but it was common practice for Navy Fleet & Seabee Divers. I thought it was brilliant, as your PQS becomes a quick study guide in itself. When you research and find the answers to those tasks and/or questions, write them in the margins of the page. This will help in two ways: by writing down the answer it will help you to memorize what you were writing and it will show the PQS qualifier that you have researched and taken the time to take notes on that particular task. This will also act as a study guide when you are preparing for an oral board or written exam. Another way to do this is to have a notebook as an answer key to the PQS/JQR, this will keep you organized in your studies and again show the PQS qualifier that you have taken the time to research the tasks. This alone will show to the PQS qualifier your commitment and dedication to your job. As a non-qualified team member it is imperative that you stay motivated and show your dedication to your superiors or they will lose confidence in you and stop giving you the time of day. Once they have lost confidence in you it will be hard to regain that back and you will have to work 10X's more than you initially would have to regain their trust.

There are two types of people: the ones who make things happen and the ones who watch. You have to be a go-getter to win over the trust and confidence of your unit and team. Your qualifications will not be easy and will take some time. If you're getting underway, don't waste an opportunity to complete a task you're trying to accomplish. This isn't paid joyride time. Need to steer a compass course? Ask on the way out, on the way back, or when returning from a SAR case. I kept a ziploc bag in my drysuit pocket with tasks I needed to accomplish that day, and I always asked for the opportunity and had a plan. Once we got back and we had time - the same day - I would ask for that signature while it was fresh in their minds. I used this method numerous times throughout my career and it works. It's imperative that you don't waste opportunities for learning and asking questions. Your qualifications are your lifeline. You have to be relentless in your pursuit. You will not be taken seriously and you will not be an active member of your unit until you are qualified. Be relentless. You are not a guest here. You are a member of a team with a life-or-death mission, and you must do your part.

The Coast Guard is an organization where we don't have the luxury of time and extra personnel to micromanage everyone, so you can either be a go-getter and step up to the plate and do what you need to do, or you are merely taking up space that could be reserved for another go-getter. The Coast Guard is a meritocracy. Your career will advance based on your performance, not on seniority. Leadership will notice the go-getters and they will be the ones who get the opportunities. If you find yourself in a position where you have come to terms with not being a go-getter and not placing your qualifications as a priority - do not reenlist, do not pass go and collect another 200, and please find another line of work. There are plenty of organizations in the world that openly accept mediocrity - but for my sake, the organization's sake, and yours - do not bring this attitude and work ethic into my Coast Guard. The stakes are too high.

Reality Check: If you are not qualified, you are not just falling behind - you are dragging everyone else with you. Your delay forces the unit to double-man watches, burn out your peers, and waste leadership's time babysitting tasks you should already own. One person's lack of readiness corrodes trust across the deck plate and marks you as dead weight. In this service, dead weight doesn't get carried for long.

GOALS

You have to visualize yourself achieving that goal and see yourself as one small step away from attaining it. Lots of people suffer from self doubt and self worth issues and this negative mindset will only prevent you from achieving your goals. There is a study that I reference often to my team when speaking about visualization. There were three groups of basketball players. The first group went to the basketball court and actually shot free throws to see who would improve. The second group of players sat in a classroom visualizing themselves shooting the perfect free-throw over and over. The third group, the control group, did nothing. You would think that the group who actually shot basketballs on the court would have improved - but the players who actually improved were the ones who sat in a classroom visualizing in their mind shooting the perfect free-throw over and over, and achieving that success. I use this reference to show that if you can see yourself doing it and believe, then you can actually improve in a skill and achieve a goal.

Personally, when getting qualified in a particularly hard skill (Mixed Gas Diving Supervisor), I found myself struggling to remember all the things I had to perform when it was time for me to perform - in front of the technical masters (Master Divers) who had all the experience that I didn't. This isn't just about a qualification; this is about being able to perform under pressure, in a high-stakes environment. Time is always limited, everyone is always busy, and when it was my turn to do my "drill" and prove that I know what I am doing, I needed to find a way to shine. It's unrealistic to ask people to go out daily and rehearse with you as if it were the real thing, so I used the visualization method. After work, I would sit in a quiet room and rehearse. I even drew out a rudimentary sketch of the actual setting (console, divers, comms, etc.) to provide further visualization. I acted out scenarios in my mind and rehearsed what I would do, what I would say, and how I would perform. As silly as this may sound with me writing it, it actually works. A lot of the things we are trained to react to have a "song and dance" about them. It's a rehearsed reaction to some event. You say certain things and do certain things, and that becomes your training and response - and when the real thing happens, you do what you're trained to do. As I often quoted, "You don't rise to the occasion, you fall to your level of training."

In the military, "drills" test us and train us to react to likely scenarios. When those scenarios occur, we don't improvise - we execute rehearsed responses. While our responses may not be 100% perfect, it gives us a way to counter some probable scenarios when things go wrong. This could be something such as an engine failure on a boat, a man overboard, a loss of hydraulic pressure in an aircraft, or something to that effect. It's not realistic to ask the Coast

Guard to fly an airplane just for you when you need training on this one particular issue, such as loss of hydraulic pressure, so in the absence of a multi-million dollar aircraft and an entire crew in the air flying - you can use visualization to play out these scenarios in your mind and go through what you would do in the event this actually happened. I liken this to the table top wargames that the top brass in the military might play when planning a large-scale theater operation. Tabletop exercises to point out flaws in your plans without actually executing plans.

Another saying that I like to use is, "there are two types of people - trained and untrained." A trained person has rehearsed what can go wrong, what will go wrong, and what to do in the event things do go wrong. An untrained person acts when things go wrong - often making it worse, and only sometimes getting it right by chance. Training yourself takes dedication, such as taking the time to run tabletop scenarios and do visualization drills so when it is time to perform - you're ready. You're running through that pre-rehearsed script of reactions, actions and words - and you will know exactly what to do. This goes without saying but here I am having to say it - training the wrong way will not help you at all. You have to ensure that you are training yourself with correct procedures and correct immediate actions or you will perform the wrong procedure. Perfect practice makes perfect.

Money and Finances

The Coast Guard will pay you on time, every time. You can count on it, every single 1st and 15th of the month, like clockwork. That paycheck will feel like the most money you've ever had, especially if you've been living broke before Boot Camp. It's more money than you've ever had to manage on your own, which is where the problem starts. But here's the trap: plenty of junior members blow it all on a car loan with a double-digit interest rate, a stack of credit cards, or junk they don't need. They buy a brand new truck with a 17% interest rate the week they check in to their first unit, not realizing that vehicle is a liability, not an asset. It's not about how much you make, it's about how much you keep. The Coast Guard gives you the foundation; it's up to you to be a responsible adult and build a life on it, not dig yourself into a hole.

The military makes it easy to get screwed financially. You are a target. Car dealerships outside the gates will target you. They have special "military programs" that are nothing but high-interest traps designed to prey on young, uniformed service members. Payday lenders will try to rope you in with quick cash that has an effective interest rate of over 400%, a debt spiral you will never escape. Senior members will warn you not to sign stupid contracts - listen to them. We've seen it firsthand, watched good shipmates throw their careers away over bad financial decisions. The fastest way to ruin your career is debt that forces you into NJP because you can't pay your bills or you start moonlighting to make ends meet. Financial irresponsibility is a UCMJ offense under Article 134, and the command will not hesitate to hold you accountable for it. They won't tolerate a crew member who is a walking financial liability.

If you want to stay ahead, learn the basics early. Go to your local financial readiness counselor on base and get a plan. Enroll in the Thrift Savings Plan (TSP), which is basically your retirement account. It's the best deal you'll ever get, especially with the government matching funds. Even 5% out of your check builds into real money over 20 years. That small amount now, thanks to the power of compounding, will be a life-changing amount of money by the time you retire. Build an emergency fund with at least one month of living expenses. This is non-negotiable; it's your safety net for a car repair, an unexpected flight home for a family emergency, or any other life event that will happen. Don't cosign loans, don't buy cars you can't afford, and don't fall for "easy credit." Your friend getting a high-interest loan is their problem; don't make it yours by cosigning for them. And if a salesman says, "I can get you approved for anything," run.

The Coast Guard will give you stability and a paycheck you can count on. It's a guaranteed paycheck, something many people outside the fence line will never have. Don't waste that gift. Handle your money right and you'll be free to focus on advancing in your career instead of digging out of financial holes. Your financial discipline is a direct reflection of your personal discipline. Don't let bad decisions compromise your judgment, your career, or your future.

After Action: Credibility is the real currency in the Coast Guard. You don't earn it by talking, you earn it by showing up trained, qualified, and reliable. Until you're qualified, you're dead weight. Every day you drag your feet, you burn your team, drain resources, and brand yourself as someone who can't be trusted. This service is too small for reputations to get lost - yours will follow you from unit to unit. Your first 90 days set your brand. Get qualified, stay squared away, and prove your value. Nobody's impressed by excuses.

Courses of Action:

- Prioritize quals from day one - get your PQS in hand, find a mentor, and drive the process yourself.
- Use the 90-day rule - by day 30 you should be in the PQS, by day 60 qualified on at least one station, and by day 90 trusted without supervision.
- Be relentless in learning - study outside of work, shadow qualified members, and treat every watch or duty as a chance to prove you belong.
- Protect your reputation - show up early, squared away,

and prepared. People will remember how you started, and they won't forget.

The Standard

BLUF: Standards are non-negotiable. The uniform demands excellence, and anything less should not be tolerated.

From the time I joined to the time I retired, I watched the Coast Guard become a kindler, gentler service. That, in itself, could be an entire book that traced politics, leadership, and policies that made it that way - but I'm not here for that. I think even as a society, we're getting back to a place again where we can be honest again without everyone and their grandmother being offended. Sometimes the truth hurts and that's ok - sometimes it's true. Now, I think we are getting back to where we need to be regarding military physical fitness and professional appearance and I'm not talking about politics here - I'm talking about accountability. Accountability for yourself and for your shipmates. If you get offended by what I'm about to say - close the book, walk away and go ask for a hug from someone who pities you because there is none here to give. You have two jobs, and only two jobs, as a military member: be physically fit and be mentally ready. These two things are the only way you can execute your other duties and responsibilities.

UNIFORMS

It's easy to get lazy once you get out of Boot Camp. The constant inspections and scrutiny of your uniform from your Company Commanders are over, and you think you can relax. You can't. Your uniform is your armor, your identity, and your non-verbal communication to the world. As a young E-5, I didn't pay too much attention to my uniform anymore. I lost that "Boot Camp" ethic of keeping my boots shined, my uniform was definitely not ironed and I probably looked a bit disheveled. My coworker and friend, who was an E-6 and who I looked up to and respected, mentioned to me one day about my uniform appearance. I looked down, like "what about it" - and he said something to me that I'm sure he doesn't remember, but it seared into my brain so deep that 21 years later I can still remember where I was standing when he said it. He looked at me and said, "Have some personal pride." His words stabbed me in the gut like a knife and I instantly knew he was right. From that moment on, I made it a point to be the standard in the uniform, not because it was the right thing to do, or because it made my Chief look good, or because the Uniform Manual says so - but because I realized it was a reflection of me.

The Coast Guard Uniform Regulations Manual is a foundational document for every service member. It's not just a collection of rules - it's the standard you must uphold. When you put on that uniform, you become a representative of the service, your unit, and the thousands of people who have worn it before you. A sloppy appearance isn't just a sign of laziness - it's a sign of a lack of respect for the uniform, the service, and your own role within it. We've all heard that first impressions are everything, and that is 100% true - but so are second and third, and every single day impressions. If you show up looking like a bag of hot garbage, people will not take you seriously. You may think people don't notice that your boots are not shined, or that you took your uniform right out the basket and didn't iron it, or that your unit hat is so old and you've sweat so much in it that there's a salt rim around the brim, but trust me when I say this - everyone knows. Before you say a word, before you show people how great you are and before you wow everyone with your brilliant ideas - they are looking at your professional appearance in uniform.

Being squared away is simple, but not easy. It's haircuts/hairstyles, boots shined, uniform pressed, on time, with a notebook in hand. Dirtbags are the opposite: hair out of reg, late, excuses for everything. That contrast shows up fast, and everyone knows who's who. What matters is that "The Standard" isn't cosmetic - it's written into your evals and stamped onto your name. The professional builds credibility. The dirtbag leaves a warning label.

This isn't just about looking good, it's about setting a standard for yourself and others. Think of how a State

Trooper looks in uniform. They're the epitome of squared away: razor creases, a shirt that looks custom-fit with not a wrinkle in sight, and a Smokey Bear cover pulled low in a menacing way - an appearance that demands respect. Now picture the exact opposite. Think of the stereotypical plumber - overweight, shirt stained with last night's spaghetti, unshaven, and his backside hanging out. Who's getting the respect based on appearance? (Sidenote: shoutout to the plumbers, but you know exactly what I mean.) The reality is that your appearance signals your attention to detail and your discipline. If you can't get the small things right, why should anyone trust you with the big things - like a multi-million dollar cutter, a search and rescue case, or the safety of your shipmates? We can pretend that the world is a pillow and a safe place for us to land, and that people are 'supposed to look at what's on the inside' and all that feel-good stuff - but the harsh reality is that people are hard wired to judge you, and you are hard wired to judge them, whether you like it or not.

I promise you that if you take the extra 5 minutes to iron your uniform, you will stand out. Get your haircut religiously - do not be the guy that shows up with a Wolverine neck. Do not be the girl who shows up with her hair thrown up in a bun at the last minute, looking like you just rolled out of bed, with stray hairs everywhere. You think people won't notice but trust me they do. How long does it take to grab the shoe shine kit to rub some polish on your boots and polish them with the brush? 5 minutes, max? People notice. I polished my boots until the day I retired and I can tell you with full confidence, people notice. When you put on that uniform you need to be prepared to be in the presence of anyone - your CO, the public, your Chiefs, and the list goes on. And when you show up looking like a wrinkled mess, or with salty boots that have clearly been neglected, or with a 3 week old haircut with a hairy neckline - people notice.

Professional appearance reflects professional standards. A rusty, neglected boat tells you everything about its crew; your uniform tells the same story about you. And then there are the ones who say, "I actually work in my boots - I can't help if they get dirty." You know what? You're right. You're the exception. The uniform manual should have a line just for you: "Hear ye, hear ye! If your boots get dirty while doing official Coast Guard business, you are exempt from looking professional and are granted a free pass to wear dusty, crusty, salty, paint-smeared, scuffed-up, steel-toe-showing, leather-chunk-missing, barnacle-covered boots." That's not the way the world works. Nobody cares, work harder. Buy another pair of boots. It's that simple. I've been a broke E-2, trust me, I know - but nobody cares. You simply do what you have to do, and if buying 2 pairs of boots is that, it is what it is. You'll survive.

Discipline Off the Clock

Your uniform only comes off at the end of the day. Your

reputation does not. When you took the oath, you agreed to a higher standard of conduct, 24/7, for the duration of your contract. A lot of careers end not because of work performance, but because of liberty incidents. These are the preventable mistakes that leadership sees every day. Fights at bars, DUIs, underage drinking, reckless social media, or even something as dumb as getting caught with weed - all of that will follow you forever. A DUI will strip you of your license, impact your security clearance, and earn you a guaranteed NJP. Getting caught with marijuana is a non-negotiable, non-judicial separation from the service. You will be gone, not because you were a bad petty officer, but because you were a fool off the clock.

The Coast Guard doesn't care if you were "off duty." The Uniform Code of Military Justice (UCMJ) applies to you at all times, no matter your location or duty status. If you wear the uniform, your conduct is always under a microscope. Your actions reflect on your unit, your command, and every single person who wears the Coast Guard insignia. You're held to a higher standard. A civilian who gets a DUI might get a fine and a license suspension. A Coast Guardsman will get a court-martial or NJP, loss of rank, and possibly a dishonorable discharge. That one mistake on liberty can tank your evals, kill your chances of making rank, or even kick you out of the service. One lapse in judgment can shadow your record, derail advancement, and put an otherwise promising career off course.

The sad truth is most of the people who get separated

early didn't fail at their jobs. They failed at controlling themselves off the clock. They were great on the cutter but a liability in the barracks. You will be tempted to blow off steam - and you should, but do it smart. Find a hobby, lift weights, or join a sports league. Drink less than you think, have a plan to get home, and don't put yourself in situations where you're one fight or one post away from ending your career. That one punch in a bar fight or that one post on social media with the wrong comment can trigger a command investigation that will bury you.

Discipline doesn't end when you exit the doors of the unit at the end of the day. It doesn't end when you take off the uniform. It is a fundamental part of who you are. Carry yourself like you're always being judged, because in reality, you are. Every time you leave the base, you represent the Service. Don't be the person who brings discredit upon it.

Operational Fitness

Physical fitness is not a suggestion - it's an operational requirement. Your ability to perform under stress, to handle a heavy line, to pull someone out of the water, or to sprint down a pier to get to a boat depends on your physical condition. Yes, depending on your job in the Coast Guard your physical fitness requirements will be vastly different. That doesn't mean you don't have to hold yourself to a lower standard. Imagine a Coast Guard where even half the service is fitter, stronger, and more capable - the force would be unrecognizable in its power and readiness. Starting in 2026, every service member will be required to pass the PT test regardless of rate. All I can say is - it's about time. [Note: I gladly had to edit a line about suggesting a PT test for all members when before release of this, news of the servicewide PT test came out.]

When I joined, I wasn't the gold standard of fitness. I didn't have a family that had gym memberships or went running for mental wellness, or any of the sort. It just wasn't a part of my upbringing and I never really understood the whole gym phase or the excitement of being in shape. Even once I signed my name on the dotted line, I trained for Boot Camp doing the bare minimum I thought I needed to succeed. I ran a few times at the track and did pushups and situps all day. I couldn't run for crap. Like most things in life, I thought I was better than I actually was. In my mind I could "run", but when it hurt I slowed down or stopped. I never learned to push through that pain. That was 100% on me. Even after I joined and at my first unit, I would join the slow runners in the PT test. I would finish the 1.5 mile run in about 12:30 and I rolled with the pack of people who prided themselves on finishing right before the allotted time. This was right in line with my prior mindset of absolutely hating running and avoiding it at all cost - because it hurt! Poor me, boohoo. The issue was I never actually challenged myself so I had no idea what I was capable of. I lived in the comfort zone when it came to physical fitness. Yes, I could do 70+

pushups and however many situps but that was about it. Pull ups were non-existent, running was something I did because if I didn't pass I couldn't be boat-crew and coxswain qualified, and working out was something I did to pass time while I was overnight on duty. Then it all changed.

After I struck BM3 at Station Grays Harbor, all I wanted was to be a Surfman. I was hooked - the big seas in surf training, SAR calls in the middle of the night, 20-foot breakers, sideways rain. That was the adrenaline and purpose I joined for.

Then the detailer called. I'd already peeked at billet openings in Direct Access and, being a young married guy stuck in rural Washington, I convinced myself the right move for my family was back home in New Orleans. I asked for the new Maritime Safety and Security Team (MSST 91112). Orders came quick. My unit - who had trained me, invested in me, and put me in the Surfman pipeline - pretty much turned on me. I was one tow away from 47' MLB coxswain, with orders to Motor Lifeboat School in Cape Disappointment, on the path to Surfman. I dropped it all for a tactical unit that didn't exist yet.

Here's the truth: I had no idea what I was getting into. Post-9/11, the DHS Coast Guard was standing up 13 MSSTs to "Detect, Deter, and Interdict." It sounded exciting, and I didn't want to miss it. I started training - running, lifting, pushing my weak spots. Running sucked, but pity time was over.

Fast forward: Initial Stand-Up Training at Camp Lejeune, then deployment to Guantanamo Bay. Mid-watch patrols, detainee movements at night, and working with Marine Corps FAST companies. Nights fueled by mid-rats, days wasted sleeping, with my fitness sliding. Then word came - our unit was starting a SCUBA team for port security/ homeland defense missions. I actually had zero interest. My LT told me to take the PT test anyway.

It was me, a couple buddies, and some Navy guy applying to BUD/S. I hit the pool, swam a few laps, and was done - gassed. I stumbled through the rest, puked on the 1.5-mile run, and failed. Hard. First PT test I'd ever failed, and it stung. I had no passion for diving, but the failure lit a fire. That became my new goal.

Back in New Orleans, we started grinding - running, swimming, pull-ups, pool work with mask, fins, snorkel, and weighted belts. PT after PT, we clawed our way stronger. Then it clicked: consistency and discipline. At Navy Dive School we got smashed - five weeks of two-a-days in the July Florida heat, exercises counts in the hundreds, and instructors in our face yelling at us to quit every day. Military SCUBA school teaches you basic diving, physics and medicine, and by week three you're breathing underwater. Less than an hour after first learning to breathe underwater you dive in the pool and swim laps on the bottom looking face down - and out of nowhere instructors swim down and violently rip your mask off and regulator out your mouth, shut off your air, knot your hoses behind your back, and dare you to panic while you attempt to solve the problem. A lot do. The attrition rate is high but we didn't break. That course unlocked something we didn't know we had. It was pain, pressure, and doubt - but pushing through it rewired us forever. [Note: Prior to the CGC Healy dive accident in 2006 when LT Jessica Hill and BM2 Stephen Duque died in a preventable accident, most CG Divers attended Navy SCUBA course only. After, those left in the community attended the 5-month Second Class Diver Course (hardhat diving, underwater welding, etc.) and that became the standard with the creation of the Dive Lockers and later the DV rating. Later, I attended First Class Diver course (advanced medicine, closed-circuit diving, advanced hyperbarics, mixed gas, etc.]

What I took from it: you can train out your weaknesses. I was terrible at swimming and running. I got better because I had to. And in that grind, something changed. I stopped fighting the pain and leaned into it. I realized I could take it, carry it, even use it. That switch made me stronger. After that first failure in Gitmo, I never failed another PT test. Sick, hungover, injured, didn't matter - failure wasn't an option. It would've meant letting down the team, and that wasn't happening.

I'd do it all again in a heartbeat. The lesson for you is simple: train your weaknesses until they're no longer weaknesses. That's the only way forward.

GOALS

You need a goal. My goal was pinning that SCUBA "bubble" on my chest and becoming one of the few Divers in the Coast Guard. In a service of almost 50,000 there are only about 70 Divers. I trained my ass off and learned to push through the pain. When it hurts - you keep going. When it's hard - you keep trying. When you want to quit - you simply don't. When the water is dark and scary - you suck it up and get to work. When you're at your limits - you're wrong, you have more in the tank. Your mind will ALWAYS quit before your body. I watched it time and time again as a dive school instructor. Our job was to push people to the limit as a test. If you quit on yourself, you will quit on your buddy and your team when it's hard. The ocean is unforgiving. When your shipmate needs you, it's not the time for anything else except getting the job done. In Military Diving, your team is everything. Your Dive Buddy is everything. You risk it all for them and they do the same for you. We train for this. When it's hard, we go harder. You need this mentality and you need it now.

The thing about training is you need to train BEFORE you need it. If you're planning on some divine intervention of superhuman strength in a time of crisis then you're smoking what I like to call "HOPIUM". I HOPE I can do this, or I HOPE I can do that. That's not how life works. When the comedian Kevin Hart was in a bad car accident, the doctors said he would have been paralyzed had it not been for his state of physical fitness and core strength. His state of physical fitness kept him from being in a wheelchair for the rest of his life. And he's a comedian. What's your excuse for not hitting the gym DAILY and doing something?! Will it hurt sometimes? Does it suck to get out of bed early and hit the gym? As Jocko Willink says, "GOOD." You need to train, plain and simple. If not, you're putting yourself and your teammates in danger. When the defecation hits the ventilation - it's too late.

Truth

The Coast Guard physical fitness standards are a joke. Do not train for these bare minimums and think you're crushing it. This is the baseline level that some think-tank style group of consultants or civil service employees, chaired by some junior officer came up with to get EVERYONE in the service to pass. The bar is so low now that you can literally row on a rowing machine. DO NOT think for one second that if you're a healthy, fully capable person in the Coast Guard that just because you can do this you are reaching your full potential.

The Coast Guard routinely works with partner forces - DOD, federal, and international. When we work alongside them, we should be able to hold our own. Iron sharpens iron. DO HARD SHIT. Be with people who push your boundaries. If you stay in the back of the pack like I once did, you will never know your full potential. Just because it's hard doesn't mean you need to stop or slow down. Read a David Goggins book and recognize your full potential if you need a swift kick in the ass. But DO NOT accept the low bar that this service has set for its people. I love the Coast Guard but we can do better. ESPECIALLY if you're in a leadership role. As a Chief, I always put out max effort on the PT test. If Chief is slacking - so can I. If Chief didn't try to max out pushups - why should I? Not on my watch. Be the example and set the bar high for yourself - your team is always watching.

It will get easier and easier to skip the gym as time goes on. You WILL get busier the longer you stay in, and in life, generally speaking. You have to make the gym a part of your routine. It's better for your mental, better for your body, and better for your health. There are plenty of workout apps, routines, and plans that will take you to whatever your goal is. The goal doesn't matter. Staying active and staying in shape matters. Dare I mention food, but the older you get the more you need to workout, as your metabolism ultimately slows down. There is a poem that I love, "Do not go gentle into that good night"[1] - and this should be your mantra, as it is mine. DO NOT go down easy. Be a hard to hit target for life. The more active and healthy you are, the harder you are to hit. No Excuses, Always Ready.

---

[1] Thomas, Dylan. "Do Not Go Gentle Into That Good Night." 1951.

Reality Check: If you're training only to clear the Coast Guard's test, you're training to fail in the real world. No boarding team, no helo crew, no cutter detail cares what you scored on a rowing machine. When the call comes, the only thing that matters is whether you can push harder, fight longer, and carry more than the people next to you. Minimum standards create minimum sailors. Maximum effort creates professionals. Which one do you want to be?

Getting After It

When it comes to physical fitness, I'm going to beat this dead horse some more because it's that important. Some of you, a lot of you, are NOT getting after it. You are doing the bare minimum. You go to the gym with no plan, you scroll on your phone for way too long, you avoid hard things because they are hard, and you are perfectly comfortable in your comfort zone. THERE IS NO GROWTH IN THE COMFORT ZONE. If you want a bigger muscle, you have put strain on that muscle like it hasn't had before - or else it stays the same. If you want to learn something new - you have to study/read/listen to things you previously didn't know - or else you won't learn new things. If you want to get better at something - you have to do it over and over until you get better at it - or you won't get better. It's that simple. Some of you treat your life like this. Status quo. Doing the same thing over and over, and wondering where the results are or why things aren't changing for the better. You need to get out there and GET AFTER IT.

I still go to the gym almost daily. Actually, it's a Coast Guard gym and you just might see me there - and I will be getting after it. I've been going for over 20 years now in various units, gyms and training centers around the country. I see you. We see you. You're skating by, showing up to the gym as a place to hang out with your friends, play some basketball or volleyball, chat it up, barely break a sweat then leave. To each their own - but let's be brutally honest here - you are not getting after it. Most of you are simply not. Realistically, it's probably due to a lack of a good workout plan, any semblance of a fitness goal, or simply that your bar is too low for yourself. How do I know? My bar was too low for myself until I figured out what I was actually capable of by pushing myself and crushing my own bar. Don't let Boot Camp be some pinnacle of fitness for you. You need to get a good plan - and there are zero excuses now as there are plenty - and get after it. You cannot eat like shit all week, show up to the gym a few times here and there with a crap plan, and expect to look good in a uniform and function like an athlete when you need to and when you need it most. That's not how it works, especially the older you get. A lot of you are getting by because you're under 30 and your metabolism hasn't slowed down yet. Your poor habits will catch up to you hard unless you choose a lifestyle of fitness, personal discipline and motivation.

Coast Guard Instructors - I see you too. You should set the standard. Students look up to you. So when you teach them it's ok to play dodge ball for morning PT, what standard are you setting? When you teach them the only time you run is for the PT test, what standard are you setting? When you teach them it's ok to look like crap in a uniform or don't correct them when they're wrong - what standard are you setting? Be the standard. Be the person they remember as getting after it. Light a fire under your ass and understand the power that you have, and truly understand how impressionable your students are. They will emulate you - good or bad. Be the gold standard and set the tone. Hold yourself and your students to a higher standard and they will then hold themselves to it. It's a learned behavior. If you, as a trusted mentor, set the bar low - how low do you think they will set it after they leave? You have a responsibility to do hard shit. This isn't summer camp and it makes my blood boil when I see a class farting around at the gym at 6am on a Tuesday instead of on the grinder with sweat dripping off of them. If they wanted college, they would have gone. If they wanted summer camp, they would have joined the Peace Corps. They chose this. They likely had no idea what they were truly in for, but were 100% game for it. Only to find out that it's probably much easier than they thought. This has got to change. I'm not saying we have to crush the students daily (been on the receiving & giving end many times), crush ourselves, and not saying we need to have unit PT and treat it like Boot Camp - but as a service, we need to set the standard higher. If you are in any position of power and you are allowing this at your unit, training center, or team - set the bar higher. If you're at HQ and you have the ability to do so - stop treating people like fragile unicorns and apply some pressure. Pressure makes diamonds. People will rise to the challenge and meet the expectations if you make them - otherwise, you get status quo and dodge ball on Tuesdays. Stop farting around and get after it.

After Action: There's no 'your standard' or 'their standard' - there's only **THE** standard. You either meet it or you're below it. Bare minimum isn't enough. The Coast Guard has no room for passengers, and nobody cares what you think is 'good enough.' Every corner cut doesn't just make you look bad - it risks your shipmates and the mission. Every slip, shortcut, or ignored detail tells the crew they can't count on you. Standards exist for one reason: lives depend on them.

Courses of Action:

- Own the minimums, then exceed them - never settle for just passing; push until you set the example.
- Police yourself first - uniform, grooming, knowledge,

and performance must be locked on before you ever correct someone else.

- Respect the inspection game - it's not harassment, it's about proving you can be trusted under pressure.

- Protect the standard at all times - don't look the other way when someone cuts corners. Call it out, fix it, and make sure the team doesn't slide.

Own The Outcome

BLUF: You own every success and every failure - accountability is non-negotiable.

Initiative

Taking initiative is a crucial part of the Coast Guard's unofficial ethos of doing more with less. In a small service with a massive mission set, you'll constantly find yourself in situations where manpower is at a premium and every person counts. This means every member, regardless of rank, must be ready to step up. There are often less levels of leadership and more junior personnel in charge, leaving you with plenty of opportunity to take initiative. This is especially true at small boat stations, cutters, and isolated units where the command structure is lean by design. A small boat station might have a Chief and a handful of E-6s, but the bulk of the work falls to the E-4s and E-5s. As an E-4 or E-5, you can be in charge of an entire boat crew or boarding team, responsible for the safety and conduct of them. You're the one in charge of a multi-million dollar asset and the lives of your crew, which is a significant responsibility that requires you to be proactive and make decisions. That is a large amount of responsibility for a junior member and quite often, you will find yourself in situations where you need to make a decision. The stakes are real. The Coast Guard's mission-oriented culture means you are expected to take decisive action, not just follow orders. Being proactive about your decisions will yield better results, as opposed to always being reactive. When you're constantly reacting to problems, you're always one step behind the situation, and in our line of work, that can be the difference between a successful outcome and a disaster. On a boarding but the weather is rapidly deteriorating? Take action and do what you need to do, rather than waiting for things to turn worse and put everyone in danger. This might mean shortening the boarding, securing the scene, and getting everyone back to the safety of the cutter or boat, even if you haven't completed the full checklist. Waiting for permission from a distant command center could cost you valuable time you don't have. Working the hoist in the helicopter and you see the basket rapidly approaching the mast of the sailboat? Say something fast and start hoisting now, so the pilot can take immediate action rather than after the fact. Your situational awareness and quick, decisive communication could prevent a catastrophic collision. I can list a thousand scenarios in our line of work where taking initiative and doing something is the difference between success and failure. Whether it's a small thing, like spotting a discrepancy during a pre-mission check, or a big one, like making the call to abort a case due to changing conditions, your proactive action is essential. To take initiative on something, you have to care first - it's that simple. Initiative isn't a skill you learn from a book; it's a mindset fueled by a genuine commitment to the mission and the safety of your team. If you don't care, you let things happen to you and your team, and let the chips fall where they may. This kind of apathy is a cancer in a team environment. Speak up if that's you so I can never work with you, for you, or around you. I'm not being a jerk, I'm being honest - your lack of care will get someone hurt or killed, and I'd rather find out now than when it's too late. Often, speaking up when it's hard or not popular is the best and easiest way you can show initiative. It's easy to stay silent and go with the flow, but a true leader, regardless of rank, knows when to speak up and offer a better solution, or to point out a potential safety issue that others are ignoring. As we often do more missions with less people to do them, things inevitably fall through the cracks. This is a fact of life in the Coast Guard, and it's where initiative truly shines. Going on a boat? Look at the weather, even if you're not in charge - you're going out too. Don't assume the coxswain has seen the latest forecast; they might be busy with other pre-mission checks. You have a vested interest in your own safety and the safety of your crew, so take the initiative to be informed. Show up early and be ready to help the team with carrying gear to the helicopter - even if it's GASP - not your job! A quick reaction to a changing situation or a willingness to carry an extra bag of gear because someone else is running late shows that you're a valuable part of the team. This is a team sport and not many things happen without the team. Your individual effort contributes to the whole, and the whole is greater than the sum of its parts. The best way to be a team player is to cross-train, help out when things need to get done, and make yourself available to the team for anything that comes up. Learning a little bit about what other people do not only makes you more versatile but also gives you a deeper appreciation for their role. You don't have to wait for someone to tell you to be early and you never should. Punctuality is a sign of respect and discipline. If you're early you're on time, and if you're on time - you're late! This is a simple but critical lesson I learned early in my career. Treat time as a sign of respect. Showing up on time for a pre-brief or a mission shows your teammates that their time is valuable and you respect their commitment. If you show up late, you are disrespecting the team and showing the team that they are not important to you. It is disrespectful in my eyes. It's a clear signal that your time is more important than theirs, and that kind of arrogance will erode trust faster than anything.

The Coast Guard is a dynamic organization, and if you're not constantly learning, you're falling behind. Take initiative and when changes happen - which they will, look them up and learn them. This might be a new policy, a manual update, or a new piece of equipment. This doesn't require much effort but in the ever-changing field you work in, you need to stay up to date with the latest information. Your knowledge base is what makes you an asset, and it's your responsibility to keep it current. You don't need someone to tell you to think a few steps ahead and be ready to take action on things that need to get done. Being able to anticipate what's next is a hallmark of a professional. In all fields of work things happen like this: task-planning-logistics-work-admin. From YN to MST, this is how things get done. You should always be thinking of what comes next. Don't wait for the next step to be handed to you; be ready for it before it even arrives. I had the unique privilege of witnessing Army E-2's (Privates) in action while at Dive School. While the Army operates differently, this observation was a stark contrast to the initiative-driven culture of the Coast Guard. Army Privates do not move or do anything until they are told to. "Hey you - grab this and go over there." Then they move. When the work is done and things inevitably need to get done, like PMS the gear (post-use maintenance) - they wait until told to do so. Granted, E-2's don't know much and have no expectations of liberty or what comes next, but it was eye-opening to watch. Someone with a few years of experience knows what comes next - post-mission things need to get done. Gear needs to get cleaned, PMS needs to get done, boats & helicopters need to get washed, and post-mission admin needs to get recorded. This isn't a complex puzzle; it's a logical sequence of events. Do not be the person who stands around and acts like there is nothing to do or you have no idea what comes next. This is the surest way to get a reputation as lazy and unmotivated. Do not be the person who stands around and waits for someone to tell you what to do when it's not rocket science. A junior member who is standing around after a case is over, waiting for instructions, is a burden on the team. Always be thinking of what comes next. There are people who forget that this is a job - and like to go for rides and hang out. The thrill of the mission is real, and the satisfaction of a job well done is hard to beat, but the work isn't done until the gear is stowed and the paperwork is complete. While it is one of the greatest jobs in the world, something always needs to get done. Be the person that is thinking ahead and be ready for the next evolution. Your superiors will notice your initiative and you will gain their respect, which leads to more responsibility and freedom from micro-management. When people recognize competence, they turn down the micro-manager dial because you show that you don't need it. They trust you to get the job done right and on time. There are plenty of people who will wait to be told what to do and wonder why they are micro-managed so much. They create their own reality by showing they can't be trusted with the smallest task. Think ahead, take action, and be proactive.

Complacency

I use this analogy when explaining complacency to someone. If there is a paper ball on the floor and no one picks it up, pretty soon that piece of paper belongs there. People are ok with it, because it's "always" been there. It's part of the routine. It's a part of the floor now. It becomes part of the scenery, an acceptable imperfection in an otherwise well-maintained space. Yet, when someone new comes in they ask, "why is that piece of paper there?", people will be like "what paper - oh, that? It's just been there, no idea." This is the perfect example of how complacency normalizes a problem to the point where it becomes invisible. Don't be like that. Never be the person that is ok with the phrase I dread, almost despise - "That's the way we've always done it." This phrase is the enemy of progress and the silent killer of efficiency and safety. Pick the damn piece of paper up. It's a small act of initiative that shows you care about your environment and the standards of your team. Be the person that asks the question "why?" Why do we do it this way? Is there a better, safer, or more efficient way? This simple question can expose flawed processes that have been blindly followed for years. In today's modern Coast Guard, the technology, tools, and the way we do things is changing rapidly. I would argue, faster than any other time in history. The days of doing things one way because that's what the old Chief did are over. New ways of accomplishing the missions are coming online and the everlasting battle of old ways vs new ways is always there. This is a battle you will face at every command. There will always be the people that become comfortable with knowing what they already learned and have no interest in learning more. These people are a liability, not an asset. You have to avoid those people like the plague. They will drag you down with their apathy and resistance to change. Nothing shows a lack of care more than someone who learned something, doesn't care to learn the new way, and has no interest in bettering themselves. This isn't just about personal growth; it's about mission readiness. It is lazy, goes against the literal core value of Devotion To Duty, and has no place in the modern Coast Guard. Complacency kills

- plain and simple. This isn't a cliché; it's a hard truth written in the pages of our history. I've had people make a mistake and order the wrong equipment for us, that had we used it and not cared, it could have hurt someone. In this case, their oversight was caught by a team member who took the time to check the part number. I've seen people conducting PMS on life support equipment and not following the PMS card, because they thought they had it memorized - only to be shown they are skipping steps and important procedures when confronted about. The moment you think you know something better than the manual is the moment you introduce a fatal flaw into the system. On a beautiful clear day, I've seen people run aground because they "thought" they knew where they were in the channel, and had done it plenty of times before. Their familiarity bred a false sense of security, and their complacency led to a hard grounding. I could write a whole chapter on near-misses (almost mishaps), actual mishaps, and how complacency has killed or hurt people in the line of duty.

I had the unfortunate but honorable duty of recovering the body of Coast Guardsman ME3 Shaun Lin, who fell from a boarding ladder and drowned in James River in Virginia, while training to board the CGC Frank Drew from a small boat. This was a tragedy that shook the entire service. I came face to face with this young man underwater while recovering his body and I wouldn't dare describe what I saw, because it still haunts me. I will carry that image with me for the rest of my life. This tragic accident was caused by complacency of the training staff. He was attending a course that taught how to board ships underway, while wearing a full kit, helmet, body armor and rifle. They outfitted the students with "water wings" or PECIs, a flotation device that you wear on each side of your hips along your waist. To deploy, you are supposed to reach down with your hands and pull one or both handles upwards, causing the CO2 canister in the device to actuate and fill up the flotation. The problem was that when ME3 Lin fell off the ladder and into the water with his 60 lbs+ of kit, helmet, MK18 rifle, and other gear - there was no CO2 canister in his flotation devices. The system was there to save his life, but a critical component was missing. Someone on the training staff failed to check the students for their CO2 canisters before training. This young man lost his life because someone became complacent during training and failed to inspect the student's gear prior to. The thing about the job that we do is - even the simplest training events can become deadly because the margin of error in/around/over the water is so small. There is no room for mistakes. All it takes is one simple mistake, one piece of gear overlooked, or one "gun-decked" piece of maintenance and you are putting yourself, your team, and others in danger. "Gun-decking" is the act of signing off on a maintenance item without actually doing it. It's a silent killer. You cannot take this lightly. You will come to hear that some policies are "written in blood" and there is a sad truth to this. The rules and procedures we follow are often the direct result of a previous tragedy. Complacency kills.

FINDING THE ANSWER

There will be many times in your career where you will have the equivalent of a tall stack of manuals that you will be required to learn. Our job is heavily dependent on specific, detailed instructions, and your success often hinges on your ability to find and apply the correct guidance. I was taught a long time ago about CYA² (CYA squared) - Cover Your Ass and Can You Articulate?. This isn't about being sneaky or dishonest; it's about being prepared and professional. This knowledge that you will be required to learn will help you to articulate the decisions that you make and cover your ass by making appropriate decisions that are in line with policy. Your ability to justify your actions with policy is your ultimate protection. If the decisions that you make are not "IAW" (in accordance with) policy - then you, my friend, are operating outside of policy and the United States Coast Guard would likely not back your move. This could be a purchasing decision with supply and buying some unauthorized piece of equipment, this could be a law enforcement decision on a boarding, or this could be a disciplinary decision for one of your subordinates. In any of these scenarios, your failure to find the right policy could land you in hot water. It is your duty to know the policies, and the only way to learn the policies is to read them and study them. You will be required to know many policies and procedures verbatim - especially some very important ones, such as authority and jurisdiction if in law enforcement, and procurement policies if in supply. You won't always have time to look up every detail in the middle of a mission, so knowing the key policies by heart is a requirement, not an option. More importantly is knowing where to look to find the correct policies. This is a skill that will serve you throughout your entire career. As a Coast Guard professional you should pride yourself in becoming an expert "Googler" and search wizard when looking at manuals. Learn to use the tools available to you. Use the directives site at USCG.mil (or internal sites) to find the manuals, policies, and instruction manuals. Use the "control + F" feature to quickly search through a manual to find the keywords that will narrow your search topic down to the answer. As an example, if I want to find if something is authorized during official travel such as laundry, I would search the directives to find the Joint Travel Regulation (JTR) and use the "control+F" feature and search for "laundry", this will bring up every instance of the word laundry to zero-in on your topic which should yield your answer. If you master this ability, you will become the go-to person for your peers and your superiors when they need the correct answer quickly. I cannot stress enough the importance of this stupidly simple and very useful skill. It separates the professionals from the amateurs. [Sidenote: from the time I wrote this a few years ago as a concept till now - ai came online. I'm sure there will be some CG ai policy search at some point - zero excuses for not knowing the answer now.] The technology is evolving to make it even easier to find the right answer, so there's no excuse for ignorance. There will be times where not knowing the correct answer can be detrimental to your career, your team or the mission. You could get into legal trouble, you could endanger your crew, or you could fail the mission. Do not accept "I don't know" as the final answer - ever. This mindset is lazy and should not be tolerated in today's Coast Guard. This isn't a college class where you can skate by without knowing the material. You will experience people throughout your career that will often be too lazy to find the correct answer and will choose the easy button on asking other people what the correct answer is, or being so lazy as to blindly following what someone else does. Do your best to guide them to finding the answer themselves because giving them the answer will only enable them, and you will soon find you become the go-to for answers easily found. You're not helping them by doing their work for them; you're creating a crutch. Why am I even talking about finding the right answers? Because as a modern Coast Guard professional it is not only about having the answer, it is about having the RIGHT answer. You will find people throughout your career that like to throw out "tribal knowledge". Tribal knowledge is dangerous. It's the kind of information that's passed down from person to person without any official verification, and it's almost always wrong. Tribal knowledge starts with sentences like "the way we used to do it is," "well I was taught like," "I think it's like," "the way we used to do it at my old unit was ____," or phrases to that effect. Listening to someone like this can lead to some sticky situations and lead you down the wrong road when you make decisions based on incorrect information. The best decisions are made with the most accurate and correct information. That means keeping up with changes in the manuals to know the most current directives or policies. In the past, there have been major revisions to some of the manuals that govern my particular rating. If I was to rely on old information and not use the current updates to these manuals (i.e. using old dive tables with incorrect decompression stop times) I am unnecessarily putting my team in danger, literally. My team's lives were on the line, and the only way to protect them was to use the most current, correct information. Tribal knowledge is that knowledge that is passed down from one member to another and taken as fact and is something like playing the telephone game where one person is told a message and by the time the 20th person is relayed the final message, the words are completely different. Do not rely on tribal knowledge at all. This is not to say there are not good lessons that are passed down, because we can most certainly learn from other people's past experiences, however, when we are relying on tribal knowledge to make important decisions that can affect our people - we owe it to our people to have the most up-to-date information to make decisions with. There's a big difference between a lesson learned from experience and an unverified procedural shortcut. It takes no more than 20 minutes to log on to find the correct manual, use your newfound "control+F" wizardry and find the answer you were looking for. Maybe even print one page to show your superiors or your peers and move on knowing you have made a decision based on current factual information. The person who doesn't take time to do this is simply being lazy and there is no other way to say it.

Everyone has a boss, and it is our duty to present our bosses with up-to-date correct information. If you want them to protect you, this is how you protect them. You give them the tools they need to stand up for you. If you make a decision based on old policies or bad information, and your bosses sign off on that information and that decision - you have now put them in a situation where they trusted you and you have failed them. Everyone is accountable to someone. They will now scrutinize the documents you send them with a fine tooth comb because you made them look stupid to their bosses. Your carelessness will damage their trust and your reputation. Mistakes do happen and are inevitable but you can avoid most of these mistakes by simply taking the time to find the correct information.

Sidenote: the very morning I wrote the passage about Tribal Knowledge, I went to work and witnessed a medical injury drill where we simulated a large laceration and had to stop the bleeding. One member put a tourniquet on as the step 1 to stop the bleeding, when other steps should've been taken first. In the debrief of the drill, the member stated "well that's how we did for the last 8 years at my old unit." I walked to the bookshelf, grabbed the Brady's Emergency Care book, used the manual-mode "CTRL+F" aka the index in the back, and read the paragraph about tourniquets aloud. This quickly disputed any issues about where and when to apply a tourniquet. Tribal knowledge reared its head and had this been a real emergency, the first responder would have done the wrong thing. Tribal knowledge is dangerous.

COMMUNICATION

Early in my career, I had a big problem with communication. My communication style: I didn't. This wasn't because I was a bad person; it was because I thought I was being a good team player by not bothering my chain of command with the details. This was a critical mistake. I always left my leaders wondering what I was up to, wondering what I was working on, or wondering what my plans were - because I kept it to myself. This caused big problems for me. My intentions were good, but the results were disastrous. If the people who are responsible didn't know what's going on with me or what I'm doing, how could they effectively lead me? They can't, and they shouldn't be expected to. I thought my actions would speak for themselves but clearly they didn't. In a busy unit, your boss has 20 other people and a dozen high-priority tasks to worry about. They don't have time to be a mind reader. I was seeing things through the view of my lenses instead of putting myself in their shoes and seeing things from their vantage point. To them - I was one of many, randomly doing things, doing who knows what, but not keeping them in the loop. I would send an email to try to make things happen and sometimes these emails would make it back to my boss and they would feel blindsided. In my mind, I was doing them a favor by not bothering them with things that they didn't need to be concerned with. I was trying to show initiative, but I was doing it in a way that made my leadership look out of touch. In my mind, I was getting things done to bring back solutions to them once I was done putting my master plan in action. To them, I was cutting them out of decisions and trying to go around them. In reality, there were no ill intentions on my part whatsoever, and I truly thought I was doing the right thing - but clearly there was a better way. This led to my communication philosophy of, "You can almost never over communicate - but you can almost always under communicate". It's a simple rule, but it saved me from many headaches later in my career. It doesn't take much to keep your boss in the loop. A quick email, text, conversation or whatever. But if they find out something AFTER the fact, and they feel you should have told them in the first place, it's a bad look for you. It shows a lack of respect and professionalism. It doesn't take much to CC in an email. This is a heads up and keeps them in the loop. It's the easiest way to prevent a blindside. This is especially helpful and considerate if you are communicating with someone above them in the chain of command (XO, CO, etc.), or from another unit (Forcecom, HQ, etc.). The moment you reach out to someone outside of your chain of command or to another unit, you might as well be crowned the designated representative for your unit, and if you're asking questions or asking for things that your superiors don't know about - standby to standby. You're no longer just representing yourself; you're representing your entire unit. In these instances, use your chain of command until the well runs dry, then move outward only if you have to and with great caution. If you have the authority to reach out to your bosses superiors or to external units, CC your boss to keep them in the loop. Again - you CAN undercommunicate, but you can almost never over communicate. Your boss may not respond, but trust me, they read it. The worst thing in the world for your boss is for them to get asked a question about what you're doing and for them to not have a clue. This is the blindside that you don't want to happen. I would go so far as to say that you should even give a verbal heads up if you're about to send an email or make a phone call outside of your chain of command. Takes 2 mins - "Heads up, I'm emailing XYZ about that thing we talked about." "Sounds good" or "Hold off on that, change of plans." Things may have changed and now *you're* not in the know (happens very often because - let's be honest here, we're not in the loop are we?) Another important thing to remember about emails - they can and will be held against you. The digital footprint you create is permanent. You better think twice about sending that emotionally charged email where you're feeling high and mighty and your ego is saying FULL SEND. Don't do it. It's not worth it. It will come back to haunt you when you come down from your emotional state. Not to mention that email is considered an official record and there are such things as record retention rules (up to 10 years) to make sure these things don't just go away for years to come. That heated email you send today could be read by a board five years from now. Better to hold your tongue, suck up your pride and live to fight another day. My wife always reminds me when I'm ranting about an email or Teams message that ignites my inner fire (it burns hot sometimes) - you don't *have* to respond. Wise words that have saved me from self-destruction more than once.

Room to Grow

Do it, make it happen, then tell me what you did. This was my guidance for my team on non-essential decisions (see Type-2 decisions in the Leadership chapter). This was my way of giving them ownership and empowering them to take initiative without needing my constant approval. As a new Chief, I tried to control everything to the point where I slowed the progress of the team to a halt. I thought I was being a good leader by being involved in everything. I wanted things to be perfect so I wanted input on every little decision. This, my friends, is called micro-management. No one likes a micro-manager and yours truly was crushing it as lead micro-manager. This kills team progress faster than anything you've ever seen, as people will stop taking ownership of things, stop making decisions that they should be making, and just start coming to you for every little thing - because that's what you're asking them to do. You're creating a culture of dependency and learned helplessness. What does this do to you? Takes up ALL of your time and gets you down in the weeds on every little detail, when you should have a 10,000' view of the situation, the team, and the objectives. This will not only stress you out as you try to fulfill your duties and responsibilities as a leader but it will stress you out and take up your time because you will be doing the jobs of other people, effectively stealing their responsibility from them. This micro-management robs people of key tasks that build experience, confidence, and ability. You are not only stressing yourself by having to make every decision that you shouldn't be making, but you are stealing opportunities for growth from your people. You have to let them make decisions. They have to experience the weight of responsibility. You have to give them ownership of things or they will never grow into the leader you want them to. You may think you're helping by making these decisions for them but you are hurting them in more ways than you realize. Loosen the reins and give them the power. The critical point in this process of allowing people to make decisions on their own is to allow them to make these choices up until the point where someone may get hurt, cause damage, etc. - then you intervene. This is the difference between a learning moment and a catastrophic failure. By delegating power, you are not absolved of your duties of responsibility for the actions of your team. You should 100% step in when needed to prevent catastrophe or injury - but only at that point. When training a new Diving Supervisor, I would let them run the show from giving the first brief, to splashing divers, during emergency action drills, all the way to the debrief after. I would keep my mouth shut and NEVER intervene unless someone was making a grave mistake that would cause injury, damage, or cause some other serious infraction. If I hold their hand through every brief by correcting them on the spot for things they missed, or call out when they forget to perform some action that they should have performed, they may never learn the lessons that need to be learned the hard way. Humans learn through failure and repetition. You have to give people room to grow and enough room to allow them to fail - but to fail softly. If they care enough, they will reflect back on the mistakes they made, which you so gracefully pointed out in the DE-brief, and they will hopefully not make those mistakes again. That's growth not failure. By allowing them opportunity for growth, only then can you train them to replace you - which is the ultimate goal here. We can't stay in this Coast Guard forever (take it from me). Our job is to train our replacements and pass down the knowledge that was passed down to us. If we hoard this knowledge and don't give our people room to grow into their roles, you are hurting the entire organization as a whole. Imagine the chaos that would ensue if the entire workforce left in 1 day and everyone was a silo of information, with no knowledge passed down to junior personnel. It would be catastrophic. We have to train our people, allow them opportunities to grow, and allow them opportunities to fail (again, fail softly). Instead of micro-managing, delegate. Give them the power. If they fail, you did a poor job of communicating your intent. Give them a task and let them run with it. If they don't do it your standard, you did a poor job of communicating the standard. See how that works? Give them the opportunity to make decisions. Like I used to say - Do it, make it happen, then tell me what you did.

That's Not My Job

Inevitably you will find people in any career that have the mindset of 'that's not my job'. However, in almost any successful organization this mindset and way of thinking is counter intuitive to productivity and accomplishing the missions. The Coast Guard is no different. You can't carry everyone's workload and yours, but you can make an impact by helping when you spot a problem you know how to solve. I'm reminded of a story from the book *How Google Works*, where the CEO was not happy with the way "ad words" were working. Someone from another department caught wind of the problems and how unhappy the boss was and he decided to look into it in his spare time on the weekend. Turns out this other engineer in another department found a solution to make ads match the searches that you enter into Google. This drastically changed the advertising platform for Google making it much more effective, making the company a lot more money, and adding value to the entire company. The team member who fixed it wasn't even assigned to this particular project, nor was he tasked to find a solution to this problem. Had he said that's not my job, Google might be a different company today. In the process, that engineer solved a problem that wasn't even his to solve, resulting in triumph for both him and the company. [Fun fact: Gmail began as a side project by a single engineer - something a 'not-my-job' mindset would have killed before it started.]

During a Coast Guard High-Risk Training Course, Boat Forces Underwater Egress, where you get strapped into a boat mock-up, lifted up over the water with simulated waves and rain, dropped and flipped upside down with the boat overturning underwater while you're still strapped in, then have to escape by performing a series of steps to get out safely, one member failed the course after several attempts. They panicked and failed the scenarios and it was my job as the School Chief to debrief them and understand what's going on. This was a critical moment for this young Coast Guardsman, and for me as a leader. After some intrusive leadership and asking questions, it was clear that they were distraught with their current unit situation, and proceeded to tell me about a poor command climate to include hazing, bullying, and harassment. While I am experienced enough to know there are many sides to every story, had I said sorry that's not my job, I would be doing a disservice to that member, to the Coast Guard, and to that member's family. It took about two hours of my day to listen and respond to that incident in a way that showed someone cared, and to address the issue with higher authority. The Coast Guard is an organization about mentoring each other, and that requires to care about each other. This also takes time. If I were to say that's not my job, I'm effectively saying I don't care. If you don't care about your shipmates, your teammates, or whatever you wanna call them, you should find another organization to work in. There will be times when we have to rely on each other such as search and rescue cases, law enforcement operations, extended deployments, natural disasters and such, and we can't just walk off the job. The only option is to care enough to do the right thing for your people and for your team. It's hard to give your team the care and feeding they deserve when you have a mindset of 'it's not my job' - because it is your job at every level of the chain of command. It's like that balled up piece of paper in the corner; somehow it got there and the longer it stays there the more it belongs there. Nobody knows how it got there and if it's nobody's job to pick up the random piece of paper when they see it, it stays there forever. A mindset of 'that's not my job' is equal to a lack of initiative, a lack of discipline, lack of care for your team and is the adverse of Devotion to Duty, a core value that's a cornerstone of the Coast Guard. Don't be the person to say 'it's not my job'.

MISTAKES

Nothing kills trust faster than watching someone point fingers after a screw-up. The instant someone deflects responsibility, the crew knows they can't be relied upon when things get serious. It corrodes the very foundation of a crew. You aren't just a collection of individuals - you are a team, and accountability is the glue that holds it together. It shifts the focus from fixing the problem to assigning blame, which is a waste of time and energy. It's an immature reflex and a sign of a professional who is not ready for more responsibility. I saw units get burned because a single person refused to own their part of a failure. That one person became a liability, and a good leader will not assign a liability to a critical watch or a crucial mission. A small mistake on a line handling evolution or a search and rescue case became a complete cluster because someone was more concerned with saving face than with solving the problem. The crew will remember that cowardice. The command will, too. The fallout wasn't just punishment - it was lost trust that never came back. In a high-stress, high-stakes operational environment, you cannot afford to have a shipmate you can't rely on to have your back. Once you prove you're willing to throw a shipmate under the bus, you've proven you are a liability, not an asset.

One tool fixes this: the Debrief. It is a fundamental part of a professional operational culture. It's how we learn from our mistakes without repeating them. It doesn't have to be a formal document - it's a mindset of constant improvement. It's an honest, candid after action conversation that makes a unit better. It doesn't have to be fancy. What happened, what went right, what failed, what will be done differently. These are the four pillars of a solid debrief. It's a no-blame environment where everyone is expected to be honest about their part in the outcome. The goal is not to punish but to understand the cause and prevent recurrence. It allows the team to understand the root cause of the failure and correct it for the future. You learn in training so you don't fail in a real mission. The debrief is how you guarantee that learning happens. If you capture that after every evolution, you'll learn faster than your peers and you'll avoid repeating the same mistakes. You will become a better-rounded operator and a more trusted shipmate. Your supervisors will notice your maturity and your commitment to improving the team, not just your own performance. They will see that you have the character to be a leader, and that's how you get promoted and earn the best billets. This process is the engine of the

Coast Guard's safety culture; it's what makes us "Semper Paratus." A professional who is committed to a no-blame learning environment is a professional who is committed to ensuring everyone goes home at the end of the day

Staying Ahead of the Tech Curve

The Coast Guard you're stepping into is not the same service I joined. The missions are the same - rescue, defense, law enforcement, environmental protection - but the tools are changing faster than at any time in Coast Guard history. You can either adapt to it or be left behind. And if you're left behind, you're not just useless - you're a liability.

Unmanned systems are no longer experimental - they're operational. UAS/UAVs (unmanned drones) extend a cutter's vision far beyond the horizon. They can track a go-fast vessel for hours without putting a helo crew at risk. They can spot a migrant raft in the middle of nowhere that you'd never find with naked eyes. UAV operators will become just as critical as coxswains. Just as the DOD is creating UAS/UAV units & drone operator specialties - it's already been reported they are creating a Robotics Mission Specialist rating to handle these types of missions (note: I had to edit out my previous prediction of this when the news broke before this book was finished).

Unmanned surface vehicles (USVs) are already

patrolling harbors and waterways. Think of them as small boats with no crew, able to sweep lanes, monitor chokepoints, or shadow high-value targets without fatigue. ROVs (remotely operated vehicles) dive where it's too dangerous for humans - inspecting hulls, checking for contraband, or surveying wrecks after hurricanes. The day is coming when the best "Diver" in your unit isn't a person, it's a machine, and your job will be to operate and interpret what it brings back.

If you're the type to say, "I don't do drones" - you're the same kind of sailor who once said, "I don't trust GPS." They got left behind. New gear isn't just in the sky or water - it's on you. Wearable tech is already showing up that tracks hydration, fatigue, and vitals. It sounds invasive, and some old-timers will hate it. But when command wants to know if their boat crew can survive another SAR case after pulling an all-nighter, or if a boarding team is showing stress levels that signal bad judgment, this data will matter. If you can't operate it, interpret it, and trust it, you'll be the weak link.

The Coast Guard is a maritime intelligence service as much as a lifesaving one, and technology is supercharging that. ISR (intelligence, surveillance, reconnaissance) used to sound like big-deck Navy business. Now it's on cutters, small boats, and even portable kits on the pier.

The future is AI-driven analysis. There is software that takes satellite imagery, radar returns, AIS (Automation Identification System) anomalies, and undersea infrastructure overlays, then flags the one vessel that doesn't belong, or flags the vessel that "went dark" on AIS . UAVs already in the Coast Guard's arsenal have advanced imaging and detection capability - to autonomously detect, locate, track and classify movement and vessels. That's not science fiction - it's already happening! SAR planning tools are being fed into AI that can suggest search grids based on real-time weather, drift models, and historic patterns before your boat even launches. The operator who can run that system will be worth their weight in gold. The operator who refuses to learn it will be worthless.

If you think your old marine VHF and a pair of binoculars are all you'll ever need, you're wrong. Next-gen secure comms, FLIRs, and sonars are reshaping how we fight, search, and patrol. FLIR turns night into day, and it's no longer just for helos - small boats, cutters, and UAVs are carrying it. Advanced sonars are mapping wrecks, finding bodies, and detecting subsurface threats faster than ever. That takes operators who know what they're looking at, not just button-pushers. Satellite comms are becoming seamless. The days of "radio silence because we're out of range" are ending. Units will be tracked, logged, and connected at all times. That transparency will demand discipline.

Case Examples: Tech in Action

Hurricane response: Hurricane Ian, 2022, Coast Guard drones flew dozens of sorties, transforming damage assessments and interior search operations - giving responders a tactical edge they didn't have before.

Drug interdictions: Operation Pacific Viper, 2025. Cutters are using UAVs with AI-enabled sensors to interdict traffickers for hundreds of miles, vectoring in boats and helos at the perfect time, resulting in the Coast Guard's largest drug seizures ever.

Port security: MSRT, MSST, & Dive units have long since integrated ROVs and sonars to inspect hulls and piers for threats. What used to take us Divers hours under miserable conditions can now be done in minutes, and safer. The effectiveness will only increase as technology evolves. These aren't "maybe someday" stories. They're already happening.

Here's the bottom line: don't resist technology. Learn it early. Get hands-on with the new systems, master them, then teach others. Every generation of Coast Guardsman has had to adapt - from sails to steam, from sextants to GPS. Survival didn't go to the strongest or the smartest, but to those who adapted fastest. I built my career by leaning into new tech as it arrived, and it paid off.

If you want to lead in the Coast Guard of the future, this is not optional. The rescues, the busts, the patrols, the disaster response - every mission will be layered with new tech sitting on top of old seamanship. Show that you can learn it, use it, and apply it under stress. That's the difference between being the future-proof professional everyone trusts, or the fossil everyone works around.

ADVANCEMENTS

The requirements of enlisted advancements differ depending on what rate you are, but they all have one thing in common - they all take time. The journey is a marathon, not a sprint, and it requires consistent effort and dedication. Some people's goals are to climb the ranks as fast as they can, and some are to take it slow. Whatever your personal goals are, everyone has the same opportunities for advancements in the Coast Guard. The fact is, it's a level playing field for people of every race, gender, or background - the only thing that matters is you. What sets people apart for advancement is simple: how well they perform their duties, how committed they are to learning the trade, how driven they are to succeed, and how motivated they are to reach the next rank. Sure, there's always some luck - being in the right place at the right time, picking up points from awards - but you can't bet on that. Assume you'll get none and carry the load yourself. Your goals of advancement may change over time as well, and that's okay. The service needs people of all ranks at all times, and it's perfectly okay to perform your duties in your current rank if that's what makes you happy. However, if you stall out for too long there will be consequences, so the name of the game in the long run is move up or move out. In a small service like the Coast Guard, the "up or out" pressure is real; we have to make room for the next generation, and that requires upward advancement. My own career is a case study in how personal goals collide with service reality. My plan was to make Master Chief Boatswain's Mate as fast as I could. I joined to drive boats, conduct Search and Rescue, and do Maritime Law Enforcement. While I loved that work, life had other plans. I discovered the world of military diving - a community I didn't even know existed - and it became my calling. This choice had immediate consequences. As a BM1 on full-time dive duty before the Diver (DV) rating was formally established, my advancement stalled. My test scores would place me high on the BMC list, but then the final calculation would hit. Lacking the sea time points my peers were earning, my name would plummet hundreds of spots. It was a trade-off I made willingly; no one forced me into those special assignments. I used to joke that I needed points for "under-sea time" to stand a chance.

Even after the Coast Guard created the DV rating and I was advanced to DVC, I faced a new kind of barrier. My performance earned me the number one spot on the Senior Chief (E-8/DVCS) advancement list two years in a row but as luck would have it, the service created zero Senior Chief Divers during those two years. Lucky me. In a tiny rating with only three Senior Chiefs and two Master Chiefs in the entire organization, there was simply nowhere to go. My choices defined my career path, but they gave me an epic journey I wouldn't trade for anything. While my rating had nowhere to go back then, I'm glad to see it's grown since my retirement: the Diver (DV) rate now includes six Senior Chiefs (E-8) and three Master Chiefs (E-9). Enough about me.

The Advancement System

The advancement system in the Coast Guard is designed to select the most qualified individuals for promotion by evaluating them against their peers. While the fundamental structure has been in place for years, recent updates have modernized the process, particularly for those aspiring to the highest senior enlisted ranks. The journey to advancement is cyclical and centers on a competitive system that culminates in a final score, ranking candidates for a limited number of vacancies.

The advancement system is a formula on paper, but it's a competition in reality. You can know the rules, have perfect marks, and ace the SWE - so can the person next to you. That final multiple score is the only thing that matters, and every single point is a battle. The points for sea time, surf duty, and awards are not bonuses - they are the tie-breakers that separate you from your peers. They represent a career of operational commitment - not just good test-taking. Don't complain about the math - find a way to earn the points others won't. I was number one on the Senior Chief list for two straight years and never advanced because the vacancies simply didn't exist. That's the cold truth: a high score guarantees nothing but a spot on a list. The service's needs come first, always. The phrase 'up or out' isn't a suggestion - it's a structural necessity. Your career is your responsibility. Master your craft, chase the tough billets, and make your final number untouchable.

Here is a comprehensive outline of the current process for enlisted advancements in the U.S. Coast Guard. While the process may change again in the future, below is the current process at the time of this writing.

1. Foundational Eligibility

Before an enlisted member can even compete for advancement, they must meet a stringent set of foundational eligibility requirements by a specific Service-Wide Eligibility Date (SED). These prerequisites ensure that only qualified and recommended individuals are considered for advancement. Key requirements include:

- **Time in Service (TIS) and Time in Grade (TIG):** Members must have served a minimum number of years in the Coast Guard and in their current paygrade.

- **Commanding Officer's (CO) Recommendation:** A crucial element, the CO must formally recommend the member for advancement on their Enlisted Evaluation Report (EER). A negative recommendation will remove a member from consideration.

- **Performance Qualification:** Members must complete all required Rating Performance Qualifications (RPQs) and Enlisted Performance Qualifications (EPQs) for their current paygrade. This is managed through the Enlisted Ratings

Advancement Training System (ERATS).

- **Body Composition Standards:**

Candidates must be in compliance with the Coast Guard's weight and body fat standards.

- **Service-Specific Requirements:**

Completion of any required leadership courses, qualifications or certifications for the next paygrade.

2. The Final Multiple Score: The Core of Competition (E-5 to E-8)

For advancement to paygrades E-5 (Petty Officer Second Class) through E-8 (Senior Chief Petty Officer), the Coast Guard uses a "Final Multiple Score" (FMS) to rank candidates. This composite score is calculated from several weighted factors, with a maximum possible score of 200 points for active duty members. The components are:

The Final Multiple Score

Component	Maximum Points	Description
Service Wide Examination (SWE)	80	A standardized test designed to assess a member's knowledge of their specific rating and required general military knowledge.
Performance Factor (Marks)	50	A score derived from the member's Enlisted Evaluation Reports (EERs) over a specific period.
Time in Service (TIS)	20	Points awarded for the total length of military service.
Time in Grade (TIG)	10	Points awarded for time served in the member's current paygrade.
Medals and Awards	10	Points awarded for specific military decorations and awards.
Sea/Surf Duty Points	30	Points awarded for time spent on cutters or at designated surf stations.
Total Maximum Points	200	The sum of all possible points.

3. The Service Wide Examination (SWE)

The SWE is a cornerstone of the advancement process for most of the enlisted ranks. These exams are held on specific dates, typically in May and November, and are administered at the member's unit. The "raw" score from the exam is converted to a standardized score, which then becomes a major part of the FMS. The content of the SWE is based on the published RPQs for that rating, and is written/edited by the Rating Knowledge Manager (RKM) of that rating.

4. Verification and Advancement Lists

After the SWE is administered and all final multiple scores are calculated, the Coast Guard Personnel Service Center (PSC) creates a rank-ordered advancement eligibility list for each rating. This list is published, and as vacancies become available in a specific rating and paygrade, members are advanced in order from the top of the list. These advancements are officially announced through Enlisted Personnel Advancement Announcement (EPAA) messages.

5. The New Path to Senior and Master Chief

In a significant recent shift, the Coast Guard is moving away from the SWE for advancement to the senior enlisted ranks.

- **Senior Chief (E-8):** Beginning in November 2025, a Reserve Senior Chief Advancement Panel (R-SCAP) will be used for the Reserve component. This panel will review a candidate's record, focusing on professionalism, leadership, and performance, instead of a written exam. The promotion process for active duty Senior Chiefs continues to be based on the servicewide exam but this may be a path for a similar change in active duty.

- **Master Chief (E-9):** The process for advancement to Master Chief now also utilizes a panel system, the Master Chief Advancement Panel (MCAP), which convenes annually to select the most qualified Senior Chiefs for promotion based on a holistic review of their career and performance. This move to a panel-based system for the highest enlisted tiers emphasizes a "whole person concept," allowing for a more qualitative assessment of a leader's readiness for greater responsibility, a method long used for officer promotions.

After Action: Nobody wants to hear excuses. You fail, you own it. Passing the blame makes you worthless. Credibility comes from accountability - period. When you screw up, admit it, fix it, and don't repeat it. Leaders trust the person who owns mistakes, not the coward who points fingers. The same rule applies to your advancement and to adapting to new technology - if you don't put in the work to stay current, you're the one holding yourself back.

Courses of Action:

- Take the hit when it's yours. Own it before someone else points it out.
- Stop making excuses - they're poison. Figure out the fix.
- Be the one who solves problems, not the one who causes them.
- Treat advancements and new systems as your responsibility - study hard, learn the new tech, and stay ahead of the curve.

The Team

BLUF: No one succeeds alone - mission success depends on trust, communication, and pulling your weight.

This is not some guide on how to be the most liked person at your unit, because I doubt I ever held that title, and that was never my intention. My intention was to serve, and to do so with integrity. The Coast Guard is an organization built on teamwork and trust, and I saw my role as a leader as being a force multiplier for both. I was there to get the mission done. Sometimes that kind of drive makes other people uncomfortable or look bad. If so, so be it - the mission came first. The more capable you are, the more capable the unit, and in turn the Coast Guard.

As a young Petty Officer, I became aware of these so-called office "politics." I saw how others gained favor by using them and swore I'd never play those games. To me, politics meant cheating the system. I thought if I mastered my craft, volunteered for the hard jobs, put my unit and the mission first, and let my work speak for itself, I wouldn't have to advocate for myself - opportunities would just come. I was close, but wrong. That's not how the Coast Guard works, and it's not how the world works.

I was never good at building relationships with a wide group. I preferred a small, close-knit circle. That was fine with me, but it came with a cost. I noticed that the people who connected with admin got better service. Not because admin was playing favorites, but because when you know someone you don't mind helping them out when they need it. It was the same with supply. They handled hundreds of requests, and when one came from someone they knew and trusted, it was easier to prioritize. People who weren't afraid to interact with command ended up with the best schools. Command's job is to make informed decisions. When school slots came open, they sent the people they knew - those who showed initiative, transparency, and engagement.

For a long time, I was taken aback by this. I had been taught that being a humble, silent professional was the way. But I realized that while the "silent professional" mindset has value, it isn't a complete playbook. In fact, it can be a shield for staying inside your comfort zone. I mocked those who mingled with command, called them brown-nosers, and prided myself on letting my results speak for themselves. What I didn't realize was that while I was keeping my head down, I was shutting myself out of the real system that drives unit success -relationships.

You don't get the mission done alone. You need the engineers to maintain the boats, the armory to maintain the weapons, electricians to keep the systems running, supply to keep things in stock, and much, much more. What took me far too long to learn is that building relationships across divisions, departments, and ranks isn't cheating or crossing a professional line. It's what makes the whole machine work. Relationships are not politics, they're the lifelines that make the Coast Guard go.

In the Coast Guard, our rank structure and relationships are more relaxed than other services. This unique culture is a massive strength. While I saw E-5s hanging out with young LTs and E-6s becoming friends with CWOs, I held myself apart. I couldn't see past their rank and refused to treat them like people. For my first few years, I was in "super-military" mode, and it did me no favors.

I was too rigid for the Coast Guard. My decisions were black or white - in policy or not, with no deviation. I'd stand at parade rest in front of a Chief and avoided any form of "fraternization" with superiors. These are the behaviors of someone afraid of making mistakes and who doesn't trust the people they work with. I probably would have made a better Marine back then; their culture is built on a rigid command structure, discipline, and respect for rank. The Coast Guard is built on a foundation of trust and collaboration.

I learned the hard way that building relationships was the only way to get anything done. It's how you get the gear, the missions, and the schools. You have to put yourself out there and advertise your brand, so to speak - so they know you. When a challenging mission comes down or an exciting training opportunity comes along, they know who to call. Squeaky wheel gets the grease in more ways than one.

Later, this became my superpower. People started wondering how my division's budget was growing while theirs was getting cut. They wanted to know how my team got so many awards, how my team was going on trips and schools, and how we were getting all of the best gear approved. The answer was simple: relationships.

I'm not talking about some transactional, give-and-take deal. People see right through that and will cut you off. I'm talking about building genuine relationships. This is the key to getting things done. If you master it, you'll navigate the unit with ease, helping not just your team but others as well. When someone needs a reliable person to help them, they will call you. When you need something from them, they'll gladly help you while others are stuck at the bottom of the list.

A fellow Chief Diver at a large Aviation training unit that we, Divers, were attached to, was astonished at how I could get things done there. He asked me, "How? I thought they hated us here?" I taught him 'my method'. "It's chess, not checkers," I told him. It became our mantra as we continually navigated the system and made it work in our favor time and time again. It's really simple at the end of the day. You have to genuinely care about people. You can't just pretend; people can tell. You must truly be interested in what they do, who they are, and how you can help them. To credit Jocko, "People can smell intentions", and it's so true. If you need something from supply and you go in there only when you need something, you're a needy jerk who is using them. If the only time you talk to the training officer is when you want some school or you need some curriculum change, it's clear that it's a one-way relationship. If you fail to make relationships with the admin staff, when your pay is jacked up or you need those orders to schedule your household goods - you're just another number. If you need something from someone and you don't have a relationship beforehand, it's already too late. Now this is not some 100% of the time rule, and people will do their job to help you because that's their job, but they will likely not go the extra mile or do you any favors in the process. They will likely not go above and beyond for you, because you're just a number. You also cannot abuse this, as that becomes obvious and your intentions start to stink. Again, genuine relationships are the key. 'But I'm not good at building relationships' - neither was I.

How do you get better at something? You put your ego aside and work on it. You say hello. You ask them about the things on the wall in their office. You ask about their kids, family, hometown, etc. You get to know them. You go out of your way to drop in and say 'what's up' when you haven't seen them in a while. You ask them what's new in their world. It's not rocket science but we tend to make it difficult like that.

Some people are naturals at this. They call it "the gift of gab". If that's you, you missed your career in sales and what the hell are you doing here anyway!? For the rest of us, we have to work at it. If you haven't read the book, "How to Win Friends and Influence People", read it. It's an easy read and it blew my mind and changed my perspective on this very topic forever. I'll summarize in a few sentences but DO NOT skip this read, it's a game changer. Nobody cares about you - they care about themselves. Ask more questions. No one cares what you did this weekend - ask them what they did. Nobody cares about your snotty nose kids - ask them about theirs. Get people talking about themselves and you will then become the most interesting person to them. Funny how that works. Get people talking about their life, themselves, their things - and they think YOU are interesting. You can fight it for years like I did, thinking it's some "game" that not's worth playing, and hope people will magically see your value and come running to help you when you need it - or you can drop your pride, start building relationships, break down these "borders" that pop up in our offices, and learn how to actually get shit done. Chess, not checkers.

Conflict Resolution and Bad Leadership

You don't have to like everyone you serve with, but you do have to function with them. On a small cutter or a remote unit, your crew is your family, your neighbors, and your support network. You are stuck with these people, so you have to find a way to make it work. The Coast Guard doesn't care if your personalities clash - the mission still has to get done. Your command will not tolerate interpersonal drama that impacts operational readiness. You'll run into people who are just different from you, and that's fine. That diversity is a strength if you're professional enough to see it that way. But you'll also run into the toxic ones: the lazy, the disrespectful, the ones who try to drag the whole crew down. They're the ones who give bad advice, cut corners, and complain about everything. They are a cancer on a unit. The way you handle that determines whether you earn respect or lose it. It shows if you are part of the solution or part of the problem.

If it's just friction, like short tempers during long watches, deal with it face-to-face, in private, and move on. This is a "professional conversation." It's about the behavior, not the person. If it's real toxicity - someone undermining the mission or making life miserable for others - you don't ignore it. That behavior is a direct threat to the safety and effectiveness of the crew. Step one: address it directly and at the lowest level, not in front of the crew. Maintain their dignity and avoid a public spectacle. Step two: if it keeps happening, document it. Write down the date, time, what was said or done, and who else was present. This turns your complaint from a personal issue into a professional report. Step three: if you've tried direct and it still fails, then and only then do you run it up the chain. You show your leadership that you tried to solve the problem at the lowest level, which is a sign of maturity. Skip straight to the chain of command and you'll just look weak. Handle your own business first. It proves you are a problem-solver, not a problem-maker.

If you're a leader and you see and feel conflict - you need to "open up conflict" and address it head on. The worst thing you can do is let it fester or sweep it under the rug. It's a leader's job to manage the culture and morale of the team, and a negative attitude is like a cancer that will destroy it. It's ok to discuss problems with your team, get to the root of it and address it. It's worse to let things slowly build up pressure then things explode in one way or another. This is where your leadership and relationship skills come into play. You have to be able to have these difficult conversations with your team members. If you feel the need to ask someone to 'take a round-turn' on something - do it. If your peer is the one at fault, it's well within your rights to tell them to knock it off. Leadership happens at all levels, and sometimes it's your peers or friends that are falling into the trap. Be a good shipmate and let them know. Sometimes we need our trusted shipmates or friends to course correct us and tell us when we're off track. That's what friends do. That's what leaders do. Be self-aware if you feel this coming on, recognize this if it's you, and slap yourself out of it.

Bad leaders are a harder problem. They have rank and authority, which changes the dynamic. You'll serve under Chiefs and officers who are outstanding, and you'll serve under some who are weak, indecisive, or flat-out toxic. They will pass the buck, change their mind a dozen times, or steal credit for your work. You can't control that. What you can control is how you perform, how you protect your reputation, and how you take care of your people. Your performance must be flawless to deny them any excuse to come after you. You build a reputation as the trusted one. You protect your junior members from their nonsense. Don't let their failures become your failures. A bad leader will try to make you a scapegoat - don't let it happen. Sometimes you outlast them, sometimes you transfer, but you never let them drag you down into their nonsense. You will leave the unit with your integrity intact. If it gets truly bad, there are systems like open-door policies, Command Master Chiefs, or chaplains - but those are last resorts, not first moves. These are nuclear options, and you must be prepared to have your entire life and career scrutinized. Don't go there unless you have exhausted all other options and your documentation is solid.

Conflict Resolution Quick Rules:

- Address in private, not public.
- Focus on behavior, not personality.
- Document after a second incident.
- Escalate only after you've tried direct resolution.
- Never let emotion override professionalism.

TRUST UNDER PRESSURE

We join the service with the intent of serving on a team and being part of something greater than ourselves, though we often don't know what that truly looks or feels like. Some have already experienced that sense of camaraderie through high school or college sports, or by being part of a high-functioning team at a job. But for many of us, we came into the service wanting those things, never fully understanding what it really takes to be part of a team.

Very quickly into my career, I learned what it was all about. I had never felt the weight of responsibility like I did as a crewmember on a motor lifeboat in the raging seas of the Pacific Ocean - launching at night in howling wind and violent seas when a mariner in distress needed immediate assistance. It was no longer about me. It was about us. It was about the crew. It was about the mariners who needed us. My fear had to take a back seat, along with my wants and needs. We were all cold, we were all tired, and we were all in it together. When the call comes, you go, and there is no stopping until the job is done and the crew in distress is brought back to safety.

That means going offshore to tow in a disabled vessel with a leaky, broken rudder. It means dewatering a boat in nasty seas, towing it through the night at a crawl, fighting fatigue, fighting the sea, and fighting time. Those experiences bond humans together more deeply than any training scenario ever could. After people experience hard things together, the bond is forged, a certain level of trust is established, and sometimes that trust lasts a lifetime.

I experienced that on another level during my first dive school training. Six of us showed up in Panama City, Florida, as co-workers - some close, some not - just people from a unit attending a school together. What happened there I will remember forever. We were tested, humbled, challenged, and we suffered together. Grueling log PT, chain drags along the pier with lengths of chain weighing close to 300 pounds each, beach PT that included "vertigo drills" designed to make you puke (aka continuously rolling head over head) to "simulate vertigo," countless water PT sessions like treading water while passing around kettlebells and 45-lb plates over our heads, the many timed bay swims - including some with twin 80-lb tanks on our backs - and anything else they could do to punish, I mean prepare, us. That suffering forged a bond that I never knew could exist at that level. It built a trust that made you part of a brotherhood - a brotherhood that meant you would put yourself in harm's way to save your brothers.

In every diving evolution there is a key position: the standby diver. The standby diver sits ready on the deck of the pier or ship while the other two divers work underwater - gear on, intently listening, ready to act if the divers below need help. A standby diver's job is to free a trapped diver, recover an unconscious diver, locate a lost diver, or respond to any other emergency that arises underwater. Every standby diver knows with certainty they are potentially putting themselves in harm's way if the call comes - yet every single one is willing to make that sacrifice if their brother or sister needs them.

There is a certain comfort in knowing that someone you trust with your life is on the surface and at the ready to dive to the depths to reach you should you need them. That person is waiting, watching, and prepared to risk everything to bring you home. That kind of trust can only be earned through shared hardship - persevering through hard times, enduring the same pain, and knowing that the person next to you has the same level of fortitude that you do.

This bond cannot be faked, forced, or simulated. It's why civilian teams, no matter how talented, will never operate at the same level as military teams. There is no equivalent crucible that civilian companies can put their teams through. They can hold "offsites" and team-building events, but nothing they do will create the bond that comes from doing hard things, facing real danger, and overcoming real obstacles together as a team.

If you want a strong team, you have to make them earn it. You don't build trust sitting in a classroom talking about it. You build it in the cold, the dark, and the uncomfortable. You build it when everyone's tired, frustrated, and still pushing. Put your team through hard, realistic training that feels almost unfair. Push them to the edge, then push a little further. Make them work for every ounce of progress. That's how they learn to operate under pressure. That's how they learn that fear, pain, and fatigue can be managed. When they come out the other side together, they'll know what they're capable of - and what each other is capable of.

The most valuable training isn't about perfection; it's about exposure. You want them to experience friction before it's real. Let them fail, then demand that they recover. Let them feel the stress of chaos while they still have a safety net. When a team has been through realistic training that mimics the worst days on the job, they don't break when it happens for real. They already know what it feels like to be cold, wet, confused, and still moving. That kind of preparation burns confidence into their muscle memory. They stop asking, "Can we do this?" and start saying, "We've done worse."

If you go easy on your people during training, you're lying to them. You're telling them they're ready when they're not. The first time they face real pressure shouldn't be on the job, when lives are at stake. Soft training produces soft operators. It creates hesitation, doubt, and panic when conditions turn hostile. The teams that perform under fire are the ones who've already lived through the grind together. They trust each other because they've bled together. Hard training isn't punishment - it's insurance. It's the difference between a team that survives and a team that fails when the pressure hits.

READING THE ROOM

As you navigate this game of office "politics", something else I learned way too late was how to read the room. Knowing when to talk and when not to is a fundamental skill for any leader or junior member who wants to succeed. You need to know when to shut your damn pie hole and when to open up. Hint: opening your mouth should happen WAY less. I thought people wanted to hear my great ideas even if it would make my Chief look bad, and I thought I should be the guy in the back of the room to call people out on their bullshit - because we "speak truth to power", right? Read the friggn room idiot. When you try to call people out in front of others, you just made a sworn enemy that would rather watch your ship sink than throw you a life ring. This is likely the same person who you will need something from later, also. Likely the same person who is in charge of your evaluations. And here you go trying to make them look stupid, instead of defending and supporting them. When you make your leadership look bad, you make yourself look bad. This is in violation of Law #1 in *The 48 Laws of Power*[1], "Never Outshine the Master." Literally rule #1. Your leadership's failures are your failures, and their successes are your successes. I draw the line at safety issues that put people into harm's way unnecessarily and you should too, but thankfully those times where we need to open our mouth and raise concern about it are not often. By speaking out openly against your boss/superior/whoever, you lose valuable

---

[1] Greene, Robert. The 48 Laws of Power. New York: Viking Press, 1998.

leadership capital that's hard to gain back. It's a fragile container and once it's broken it's really hard to put back together. Ask me how I know. If you find the urge to open your mouth in a room full of other people (training, conference room, all hands, etc.) you need to W.A.I.T. - Why Am I Talking? Is what I'm about to say or ask covered somewhere else? Do I REALLY need to ask this in front of everyone? What's my REAL intention here - not the made up bullshit you tell yourself? Make me look good and catch them off-guard? Is your goal to have them answer the question so other people can know too? Whatever your reason is, it's probably not a good enough reason to open your mouth. Another great saying I live by now is "Better for people to think you're a fool, then to open your mouth and remove all doubt."

Do you really think what you will say or ask will be so enlightening and mind-blowing that people will be like, "WOW! Thank you so much for changing my mind on the spot! I would have never thought of that had you not brought it up because I'm incompetent and would have never been successful had it not been for you speaking up today! Give this person a round of applause. Wow, to be in the presence of greatness. Amazing." Wake-up. That's not how it works. If someone in the military is presenting something to you, they already have their mind made up, likely have 3 or more COAs (courses of action), and are not asking you for your input - they are merely giving you a heads up. It's highly likely nothing you will say will be so impactful that changes will be made after the fact. If you've ever seen a junior officer aide take a note when directed by a senior officer after someone asks a question, I would bet that regardless of what the outcome of that 5 minutes of research done after that note was taken - the decision remains the same. Meaning, the note was taken for show and simply to appease you, and you will have just wasted everyone's time including your own. Again, safety is one thing and feel free to voice concerns as you see fit - that's not what I'm talking about here. I'm talking about the Admiral's visit where they show up to 'get your input' - which really means telling you how it's going to be. When the dog-and-pony show starts and you feel that urge to put your hand up and run your mouth - W.A.I.T. - Why Am I Talking? Bite your lip, sit on your hands, stare at the clock, and keep your trap shut. Take it from yours truly, the reformed junior sniper from the back of the room who has been given new life by WAIT-ing.

Digital and Social Media Discipline

The Coast Guard has rules about social media, but even if you never read them, common sense should guide you. The official policy is in the Coast Guard's Media Relations Manual, but the bottom line is the UCMJ still applies to your online conduct. The internet never forgets, and one stupid post can ruin your career faster than failing a qualification board. A failed qual board is a setback you can fix. A negative post on the internet is a permanent public record that cannot be taken back, and it may be shared and scrutinized by your command. Screenshots/screen recordings travel faster than you think, and once it's out there, you can't take it back. Someone WILL see it, and inevitably someone will report it.

Don't post pictures of restricted areas, ship movements, or operations. That's a direct violation of OPSEC (Operational Security) and can compromise the safety of every person in your unit. That's OPSEC 101. But it's bigger than that. Don't trash your chain of command online. This is insubordination, and it will be handled swiftly and professionally by your leadership. Don't post racist, sexist, or political crap - not in uniform, not out of uniform, not ever. You will drag the whole service through the mud, and it'll get handled hard. You're not a civilian anymore. You represent the service 24/7, and everything you do reflects on it.

There are plenty of stories of junior members losing rank, getting reassigned, or getting booted for social media stupidity. They get an NJP, a negative eval that kills their chances of advancement, or they go to an administrative separation board and get discharged. It might feel unfair, but that's reality. You are held to a higher standard because you are entrusted with public safety and national security. When in doubt, don't post it. That simple rule will save you from 99% of all trouble. Keep your personal life off the public feed, and keep your professional life locked down. Use your privacy settings.

Social media is a tool, not a venting ground. Using it to

air personal grievances is a sign of immaturity and a lack of self-control. We've all seen the social media videos of people venting in their car about their chain of command, or using social media as a place to air out their grievances - do not fall victim to this trap because you see other people doing it and think it's the cool thing to do. Those videos are a straight shot to the CO's desk. That means an inquiry into you, your past, your everything - it will not end well for you. Don't kid yourself - they don't waste time on long investigations when you hand them the rope yourself. Use social media to connect with family, share memories, or for entertainment - not to destroy your career before it starts. Don't let your 15 minutes of going viral cost you an entire career.

Complaining

It's inevitable when you work around other people that there will be the people who love to complain. They love to grumble about anything and everything. It seems like they have nothing better to do at times. One of my colleagues called this constant talk: Sport Bitching. When you feel like you can't control your circumstances, and you're kind of a miserable person, you want other people to join in your pity party - so you start complaining. It's so hot. It's so cold. It's raining really bad and this sucks. I'm tired. I'm hungry. I'm tired of this. I'm tired of that. Misery loves company. People that like to complain want to complain with other people who like to complain. If you become a part of this group, the negativity will start to creep in and soon it will follow you everywhere you go. You'll become a certified member of the pity party and pretty soon you'll be the captain of the squad. Team Sport Bitching. You'll find the negative in every little thing that happens and you'll start to pick apart things with the lens of a naysayer. Your "can-do" attitude (if you ever had one) will go from looking at the potential ways of getting things to done, to complaining about all the ways this or that won't work and you'll have no solutions to bring to the table, just complaints.

There's a saying that we live by as leaders - don't come to leadership only with problems, but bring solutions. If you bring up a problem to your boss, you better have thought of a solution, or else you are a part of the problem. What's worse than bringing up a problem with no solution? Bringing up a problem with no solution then complaining the whole time about it! You either choose to be a part of the solution or a part of the problem - it's that simple. It's easy to sit on the sidelines and complain about what other people are doing. Being an "armchair quarterback", where you have the hindsight of not actually being in the game, but get to sit there and judge other people for their decisions or actions, is taking a coward's way out. The quickest way to get humbled is to put yourself in the hotseat and see how well (or not well) you do. Then you might not have so much to say. If you're busy in the drama club complaining, you're probably not the one stepping up and getting things done. That requires confidence, courage, humility, and initiative. Sideline talkers love to run their mouths but avoid the game - playing would expose their weaknesses, so fear keeps them on the bench.

It takes guts to step up and lead - to stand in the line of fire knowing you'll take the hit if it goes wrong. Funny thing is, the loudest complainers and drama-stirrers are never the ones out front. They're the ones who always need to be led. Leaders take action while others simply talk about what they're doing. Don't talk about it, be about it. In survival, the most important factor is the will to live. Now ask yourself - do you think the chronic complainers will have it when it counts? Do you want someone like that in your liferaft in the middle of the ocean? If all you bring is whining and talk about how we're doomed, you're dead weight - and you might find yourself off my raft.

We need people with CAN-DO attitudes, people who can be resilient, teammates who bring out the best in each other, shipmates who can take the slack for you when you need it the most, and people who you know will have your 6 when you need someone to watch it. Avoid the negative people like the plague, and if you know you're prone to negative thinking about things - keep your pie hole shut and don't do it. For the rest of you, if you see the sport bitching squad come out to play - call them out and nip it in the bud. You cannot let them start, continue, and grow in strength and volume. Misery loves company and soon your entire team or unit will be on the squad. Problems require solutions and last time I checked, complaining about things never fixed anything.

After Action: No one makes it through this service solo. You can be the hardest charger in the unit, but if you don't pull with the team, you're a liability. Being part of a crew means putting the mission and your shipmates ahead of ego. Trust is everything - it's built fast by showing up squared away and destroyed just as fast by laziness, gossip, or selfishness. The Coast Guard runs lean, and every member is a link in the chain. If you're weak, toxic, or checked out, you don't just hurt yourself - you break the team.

Courses of Action:

- Earn trust early - show up on time, in the right uniform, prepared to work, and back your shipmates without being asked.
- Kill toxicity - no gossip, no whining, no dragging others down. The team doesn't need passengers or drama.
- Carry your weight - qualify fast, master your role, and be the one people count on when things get chaotic.
- Be a multiplier - share knowledge, mentor juniors, and look for ways to make the crew stronger than the sum of its parts.

Lead The Way

BLUF: Leadership means setting the standard through your actions and earning respect long before you wear the rank.

Defining Your Style

When you become a leader you will have to define your leadership style, either through doing the wrong thing, or by studying the art of leadership. The Coast Guard is full of examples, both good and bad. You'll see leaders who are revered for their advanced knowledge, their experience and skill, and their ability to get a crew to perform beyond what they thought was possible. Conversely, you'll see leaders who are reviled for being petty, vindictive, or just plain absent. Your leadership style will be a sum of these experiences, but you have the power to consciously choose your path. For some people it comes naturally and they tend to do all the right things without ever trying. They're the unicorns, the ones who seem to have an innate sense of influence and a knack for inspiring others. These leaders are the gold standard, and they are few and far between. They're the ones you see a whole unit rally around, the ones who can turn a chaotic situation into an organized operation with a few calm words. For the rest of us, we have to study the art, learn the ways, and work hard to be a better leader.

I call it an art and not science because in science things are repeatable in an exact way. However, people are random and the same tactic or technique may not work the same twice in a row, so leadership becomes this dynamic playing field that will keep you on your toes and challenge you to your core. The same approach you use to motivate a salty Chief might fall flat with a brand-new, 19-year-old Seaman. You have to be a chameleon, constantly adapting your approach to the individual and the situation. This isn't being fake - it's about being effective. You'll be dealing with everything from a junior member whose parent just passed away, to a senior member who's pushing back on a new policy because "we've always done it this way." Your toolbox needs to be full, and you need to know which tool to use and when. When most people are put into leadership roles for the first time, they unknowingly tend to emulate a leader from their past, who in one way or the other, made an impression on them, good or bad. This is a default setting, a survival mechanism. You subconsciously fall back on what you've seen and what you know. If you had a leader who led with fear, you might find yourself doing the same thing. If you had a leader who was hands-off, you might adopt a similar style. This is why studying leadership is so critical - it allows you to break this cycle and build a style that is effective and true to your values, not just a copy of someone else's. I was in leadership positions for many years but I never took the time to really understand it. To me, it was simply being in charge, trying to make the right decision for the right reasons, making judgment calls when you had to, and carrying out the orders that were given to you.

It was always 'mission first' for me and that's how I treated the people I was in charge of. This "mission first, people later" mentality is a common trap, especially in an operational environment where the stakes are high. It's easy to see your team as a collection of assets to be deployed, rather than a group of people with their own lives and challenges. While this got the job done, it didn't do me any favors of building high-functioning teams in the process. We got the job done - but sometimes in a dysfunctional way, which doesn't make for a good time. A team that's just a collection of people doing their jobs is not a team - it's a squad of individuals. They'll get the work done, but they won't have each other's backs when the real pressure hits, they won't innovate, and they won't go the extra mile. The Coast Guard is full of missions that demand trust and teamwork, whether it's a boarding team on a vessel, a boat crew in heavy weather, or a flight crew searching for a missing person. You can't fake that level of cohesion, and a purely "mission first" approach won't build it. I didn't have a leadership style and it showed. I prided myself at getting the job done no matter the cost, and this doesn't make for a good leader or a good teammate.

I remember my most hated moment as a leader like it was yesterday, and I cringe at the thought of it. At a new unit, I was put in charge of the Diver's Safety Assessment (DSA). If your unit is operational, you'll go through some form of this -whether a Standardization (STAN) Team inspection for boat forces, aviation, or other missions. These inspections are the real test of readiness and leadership. Evaluators, usually senior members with decades of experience, are looking for any cracks in standards or procedures. Fail, and the consequences are severe: the team can recommend shutting down operations until deficiencies are fixed. That failure reflects directly on you, your team, your command, and ultimately the CO. Since the CO is responsible for operational readiness, a failed inspection doesn't just stain the unit's reputation - it can derail careers. The stakes couldn't be higher. My team was not taking this inspection seriously and we were way behind schedule. This inspection requires a certain level of showmanship: organized records, spotless spaces, squared-away gear, and a sharp presentation for the inspectors. Details matter and first impressions are everything. The inspectors are looking at everything - from the way a training record is filled out to the cleanliness of the trailers. A disorganized workspace or a missing piece of paperwork can be a red flag, signaling a lack of discipline or attention to detail. My team was not taking this seriously and I became unhinged and lost my cool. In front of everyone, I yelled, cursed, pointed fingers, spoke down to, and did all the things someone who lost their cool could do. I was furious and frustrated. This emotional outburst was a classic sign of a leader who had lost control of the situation and, more importantly, themselves. I left pissed off and mad - but still in the same position we were before I started my tantrum.

After a long, hard self-reflection of how we got into this predicament, it hit me like a ton of bricks - it was all my fault. I spent the next few hours replaying every interaction, every missed opportunity to get the team on track. The team was doing exactly what I told them to do - not much. My leadership style was to use "big boy rules" and let them run with it as they saw fit. "Big boy rules" is a convenient way to abdicate your responsibility as a leader. It's a hands-off approach that assumes your team is psychic and knows exactly what you want and when you want it. This is a fantasy. I failed to set clear expectations, build deadlines, or create an environment where questions could be asked. I never established a timeline or managed expectations - I just assumed the team knew what to do. Some did, but many didn't, and instead they worked off what they thought was right. The outcome of their process was entirely my fault. As the leader, you are the one who sets the tone and the direction for the team. The team is a direct reflection of the leader, and my reflection was a disorganized mess. After some overdue apologies, I set clear expectations, realistic deadlines, answered questions, and tracked progress. The team rallied, and in the end we succeeded - making the unit and command look good, which was the goal.

In that, I learned a valuable personal lesson about leadership that I should have already known - anything that your team does while you're in charge is your fault. Good or bad. This is a foundational principle of effective leadership. It is not about blaming your team; it is about accepting total ownership of the outcome. If they are not taking the upcoming inspection seriously, it's your fault. If they are not showing up on time, it's your fault. If one of your boat crew members accidentally discharges an M240B machine gun while escorting a US Navy nuclear submarine - your fault (ask me how I know). You have to take full accountability for your team's actions and be fully prepared to fall on the sword when the time comes. This is the difference between being a manager and being a leader. A manager assigns tasks and reports on progress. A leader owns the results, both good and bad, and takes responsibility for the team's successes and failures. If you're not ready for that level of responsibility, you don't have any business being in charge.

The best thing you can do is be honest with yourself, own up to the things you are NOT doing, and correct the course as soon as humanly possible. You can make a million excuses as to why your team is not functioning as a cohesive team, and while it may be true on some levels, you will never achieve the results you desire if you don't take full responsibility for the entire situation. Blaming a lack of resources, a difficult command, or a dysfunctional policy is a waste of time. Your job as a leader is to navigate those obstacles, not complain about them. Complaining never fixes anything - taking action does.

If you don't have a leadership style and don't know

where to begin, as cliche as this might sound due to the popularity of the book and person now, I would start with Extreme Ownership by Jocko Willink & Leif Babin. This book is not about being a Navy SEAL - it's about the core principles of leadership that apply to any high-stakes environment, and the Coast Guard is full of them. You don't have to be a SEAL to understand the concepts in the book, and I would be lying if I said this book didn't change my entire leadership style for the better. It became my super power that allowed me to lead my team under extreme pressure, very high op-tempo, and allowed us to outshine any other division and team in my unit. We were faced with impossible deadlines, staffing shortages, and high-risk situations, but because of a shared commitment to accountability, we always delivered. I was pissed I hadn't learned these lessons earlier in my career, but you can't go back. I won't quote the core principles from their book - I don't need a lawsuit - but trust me, you won't be disappointed.

Leadership is like a knife - a skill that can be honed and sharpened. Some are born with a katana, sharp and ready. Others start with a butter knife -dull, but workable. Sharpen it into a razor-edged blade, and it becomes far more than a tool to get by. A butter knife might scrape through, but only a sharp blade can perform surgery - or lead a crew through a life-or-death situation with precision and confidence. You will fail as a leader. I only pray that you fail softly and don't get anyone hurt in the process. The mistakes you make on a cutter in the Pacific are very different from the mistakes you make in an office building. Humans learn through failure - but only if you run a mental after-action, ask what you could have done better, and stay honest with yourself. Skip that, and you'll repeat the same mistakes, making both you and your team pay the price.

People First

My personal leadership philosophy became rather simple over time: People First. If there was a decision to be made, I took my team into consideration first and foremost. This doesn't mean you ignore the mission, but it does mean you recognize that your people are the engine that drives the mission. Without the individual people on your collective team - along with their knowledge, their dedication, their specialty skills, their commitment, their strengths, weaknesses and all - the team is nothing. In the Coast Guard, our people are our most valuable asset. A boat is just a piece of metal, an aircraft is just a machine - without the skilled and motivated people operating them, they are useless. The Coast Guard is a team sport - the team is everything.

Being part of a team takes dedication. Leading one takes constant attention. Your people aren't robots. They have families, goals, strengths, weaknesses, and lives outside the Coast Guard - and those lives will affect their performance. As a leader, you must be humble, respect the privilege of leadership, and treat your people as human beings. Empathy is non-negotiable. Put yourself in their shoes and remember: every decision you make has consequences.

Before you make a decision, policy, or rule - you need to think of how this impacts your people, both negatively and positively. For example, a decision to change the watch schedule might seem like a good idea for efficiency, but how does it impact the new parent who just had a baby and needs a predictable schedule? If you think there may be negative repercussions to the team with a decision or change you're considering, use your peers or trusted advisors (fellow E-5/ fellow Chief/mentor, etc.) and bounce it off of them. Maybe you're not seeing all of the sides of the problem and how it may impact the team. Asking for help isn't a sign of weakness; it's a sign of a strong leader who values different perspectives. You might think, "Nah, not me. I think it all through before I make a call." Sometimes, when you are in the weeds of a problem, you can't see the big picture clearly. Asking someone who is detached from the problem with no stake in the outcome can reveal blind spots you missed.

Having a "People First" leadership style is not easy - it demands a lot from you. It's about playing both offense (making changes yourself) and defense (speaking up for your team). Orders and policies will come down the chain of command that may affect your people in ways leadership didn't foresee - blanket policies, one-size-fits-all directives, and so on. As a "People First" leader, you must gather the facts, do your due diligence, and use the proper channels to respectfully go to bat for your people. This is where the rubber meets the road. Leading is easy when everything's going well; it's much harder to push back against a bad policy from above. Doing so takes precious time, vigilance to see it through, and sometimes the courage to speak up - but if it's the right thing, you do it because it's right.

As a leader, you are both a trusted advisor and a shield for your people. Your role is to be their voice when they have genuine concerns - and with experience, you'll learn to tell the difference. A genuine concern is a systemic problem that hurts the team's ability to do its job or takes a toll on their well-being. A gripe about standing an extra watch because someone is hurt, on leave, or away at training doesn't rise to that level. To recognize the real issues, you may have to ask the right questions and dig deeper, because some people will simply "suck it up" and press on. If you don't look past that surface toughness, you risk missing problems that quietly erode performance and morale.

Being a leader also requires constant self-improvement. You have to constantly assess and re-assess if you're meeting the standard. You have to WANT to be better, not only for yourself but for your team. You have to become a disciple of leadership. Read books, listen to podcasts, listen to audiobooks, read articles - do what you must to fill your tool bag with more tools than you think you need. Use your mentors and peers for advice when you need to and even when you think you don't need to. They likely have already dealt with some of the issues that you're currently facing, and may have another angle that you're not considering. If you have only a single direct report, you owe it to them to be the best leader you can be and not just winging it because you wear some collar device that says you're in charge. Study hard, do your homework, and always take care of your people. It's not about you.

Control Your Bubble

Look, when you join the Coast Guard, it's easy to get caught up in the vastness of the organization and its missions. You see headlines about drug busts in the Eastern Pacific, rescues from hurricane-ravaged communities, or new policy initiatives coming out of Headquarters. It's a lot to process, and it can be overwhelming trying to figure out where you fit in and how you can possibly make a difference in such a massive enterprise.

You might feel compelled to fix the systemic issues you see - like why a specific policy seems inefficient or why a certain process doesn't make sense. And don't get me wrong, having a keen eye is a good thing; it means you're paying attention. But if you're not careful, you'll find yourself trying to solve the world's problems, or trying to solve the larger Coast Guard's problems - it's easy to get overwhelmed by trying to fix everything. Senior enlisted leaders and officers, myself included, have spent entire careers trying to influence change at higher levels, and it's a slow, often frustrating process. You'll burn yourself trying to change the world before you ever get a chance to make a real impact on the things that actually matter - like your immediate work environment.

At the end of the day you need to focus on what you actually have control over - your team, yourself, your standards, your PT routine, your ability to be on time, your habits, etc. This is your "bubble." It's the most critical space you're responsible for. Whether you're a Fireman in the engine room or a BM2 on the buoy deck, the operational readiness of your unit starts with you. Is your gear clean and ready? Are you squared away for the duty day? Have you mastered the technical competencies of your job? Can your team rely on you? These are the questions you should be asking yourself. Trying to focus on high-level policies that you ultimately don't have a voice in will get you stressed out and overwhelmed in the long run.

Your energy is a finite resource. Don't waste it on things you can't influence. Instead, invest it in your immediate environment. By focusing on your "bubble" and what you actually can control, you can dominate your space and actually make impactful change that you and others can see. When you show up every single day-on time, prepared, and ready to contribute - you're not just doing your job. You're setting a standard. You're demonstrating to your shipmates that excellence is non-negotiable, and that example is contagious. That's how you build a strong unit. That's how you make a real, lasting impact in the Coast Guard.

Praise in Public, Correct in Private

One of the most effective leadership habits you can develop is knowing when to keep your mouth shut and when to make noise. It is a fundamental skill that separates the good leaders from the bad ones. The old saying "praise in public, correct in private" isn't just a cliché - it's the difference between building a team that respects you and one that quietly hopes you fail. In a small, close-knit unit, like many Coast Guard units are, every action you take is on full display. Public humiliation doesn't toughen people; it shuts them down. It is a weak tactic used by leaders to assert authority through fear, and it is a corrosive approach. If you blast someone in front of the crew, they'll remember the embarrassment, not the lesson. What you actually teach is that you can't be trusted with their dignity. You will have broken the trust required for a crew to operate safely and effectively.

Correction still has to happen - standards don't bend - but the setting matters. Failure to uphold a standard is a failure of leadership. Pull them aside after watch, walk with them back from quarters, or call them into the office. The location should be private and neutral, and your tone should be calm. Deliver the feedback bluntly but privately. Focus on the behavior and its impact on the mission. "You missed a step in the boat check, and it delayed our getting underway" is professional. "You're an idiot for forgetting the checklist" is not. This keeps the focus on fixing the problem instead of fighting off shame. The person is more likely to listen and absorb the lesson. When the person comes back squared away, the team sees the standard upheld without you ever needing to put on a public show. This earns you respect from everyone, not just the person you corrected. Note: Safety concerns are different. If there's a risk of someone getting hurt, you address it on the spot and directly - no delay, no pulling aside later.

Public praise works the opposite way. It is a powerful, underutilized tool for building morale. Recognizing someone's effort in front of their peers sets a benchmark everyone else can see. It costs nothing but a few words, and the return is immeasurable. You're signaling, "This is the behavior that earns recognition." Whether it's a sharp uniform, taking initiative on a messy task, or nailing a qualification board, a few words at the right time lift morale and reinforce culture. People will work harder when they know their effort won't go unnoticed. They will see that good work is valued by leadership and their peers.

The balance is simple: use private spaces for correction, and public spaces for recognition. This is the litmus test of a professional leader. Leaders who reverse this - ignoring good work and embarrassing people in front of the crew - burn through credibility fast. They are a liability to the command. Discipline and standards aren't just about rules; they're about how you handle people. A leader's job is to train and develop their crew. Get this right, and your team will follow you even when the work gets ugly. They will trust you with their safety and their dignity, and that is the most valuable asset you can have.

80/20 Rule

It's said that 80% of your time as a leader is spent on 20% of your people and I, personally, found that to be highly accurate. If and when you are in a leadership position, you will have some people on your team that, let's just say - need more attention than others. They will need more coaching, more guidance, will ask more questions, and will need more "care and feeding", so to speak. These people are not bad workers by any means, they just may require more from you than the rest of your team. This is a reality you must prepare for. You will have to dedicate a significant amount of your time and energy to a small portion of your team. Some people just "get it" and require little oversight, little supervision, and know how to navigate from a-to-z on a task. These people are generally self-motivated, self-starters who can operate independently and have confidence in making decisions, and often make the right decisions.

Sadly, these people often get over-looked by the busy leadership, because while the go-getters are getting after it and making things happen, the leadership is busy putting out fires with the 20% who need guidance or have found themselves in some kind of trouble. Don't allow yourself to get sucked into this leadership trap, where you forget about your high performers while you're busy coaching, mentoring, or reprimanding the people who need it. Your high performers are the ones who are quietly making the machine run, and if you neglect them, they will burn out or, worse, become a part of the 20%. The high performers usually don't need much, but everyone likes recognition and acknowledgment for their efforts - even if it's a small gesture here and there. A simple "thank you for all your hard work on that project" or a public shout-out can go a long way. The 20% of people will test your patience, make you question your leadership abilities, and if you care enough, frustrate the hell out of you sometimes.

These are the people that drag their feet when they need to get qualified by not studying, not being proactive enough, doing the bare minimum and never anything more, treating deadlines like moving targets, and generally have an "all about me" attitude. These people can sow seeds of discontent in your high-performing team if you let them, because to them, their individual needs are higher than the team's needs. Their lack of effort and poor attitude can become a cancer that spreads to the rest of the team. They will say they are a team player - but their actions speak for themselves. You have to take their actions at face value and not be swayed by their hollow words. There is generally one way to deal with these types of people on your team - hold them accountable.

If you fail to hold them accountable for missing deadlines, poor performance on their tasks, being late, or any other subpar actions, you will lose the entire team. You'll start to wonder 'what happened' when nobody cares anymore, when your high performers are leaving you out of the loop, when things start to spiral on your team and everyone is doing their own thing and there is a general sense of chaos surrounding your team. The truth is - if you give people an inch they will take a mile. This is not a cynical view of humanity; it's a realistic view of human nature. Most people will follow the path of least resistance. If you fail to hold your team accountable, you will lose the respect that you worked so hard to gain, and you may never get it back from them.

Some leaders think they are doing people a favor by not holding them accountable but in fact, the opposite is true. I've seen leaders sweep things under the rug, with perfectly good intentions of not rocking the boat or making a big deal out of something small, only to have that same person do something more egregious and ultimately force their hand on a much larger scale. If you're new and reading this, you may think that the military is like how they portray it in the movies, where leaders rule with an iron fist and your unit is run like a machine, with leaders barking orders and people snap-to like robots and do whatever they're told, immediately without question. While, yes, that is exactly how Boot Camp/ basic training is, that is nothing how the actual Coast Guard operates. The Coast Guard today is a more professional and respectful environment, and you may be surprised to find that you're often treated with consideration and respect as a member of the team. With that, some people start to take that extra mile when they're given that inch - and you wind up in the position described earlier. Even as an E-4/E-5, you will have to draw some lines in the sand and hold your peers accountable when you're in a position of leadership. If you don't, you will get walked all over and made a fool of.

I'm not telling you to be a tyrant and run your crew like a dictatorship, but you better establish some hard lines that must not be crossed without consequence. For example, if your duties involve guns and ammunition, you must demand 100% accuracy, accountability, and safety from your team. Even an E-3 or E-4 in charge will be held fully responsible. This is non-negotiable. There are countless examples like this, but the point is that as a leader you must define your core values and decide what lines you will not allow to be crossed. You also have to cross-reference these with the Coast Guard's own non-negotiable policies - their lines in the sand - and determine what you will accept and what you will hold people accountable for. Fortunately, in the Coast Guard, experience and leadership responsibilities tend to rise together, so with experience and time you'll be prepared for the decisions you'll face - and you'll get plenty of both.

To make my counter-point for this "holding your people accountable" argument, is a lesson that I learned after many years of leadership: show grace. We're not working alongside robots here. We're all just humans, with most of us trying our best under the circumstances, and we all fall short of perfection. We will all make mistakes and that's exactly how humans learn. Hopefully our mistakes aren't so serious that we hurt our team, ourselves, or others. If there is something to be learned in a problem, help people to see it by teaching them instead of punishing when appropriate - and when appropriate, show grace if you can. Not every rule that's broken is worthy of written documentation. Not every mistake that someone makes is worthy of punishment. Think back to a time when you made a mistake and you didn't get punished, reprimanded, or faced consequence for. Everyone has. So don't be so hypocritical as to punish everyone for everything when you know damn well you've likely done the same or worse.

Take this with a grain of salt and use this sparingly, as it's a tool in your toolbox - but remember that if we all got punished for every mistake we made we would have zero successes in our life, job, relationships, etc. Showing grace doesn't mean you are weak or a pushover. It means that you are able to recognize faults or problems as teachable moments and help others through teaching rather than punishment. Almost always, you can trace that problem back to yourself and your lack of leadership in some area. It's a slippery slope to say that you're perfect, so don't expect other people to be either. Hold your people accountable, but show grace when you can.

After Action: Leadership isn't about collar devices or who gets to bark orders. It's about setting the pace and owning the example. Your people will do what you do, not what you say. If you cut corners, they will too. If you show up squared away, they'll follow. The Coast Guard has no room for leaders who hide behind rank, chase popularity, or avoid hard calls. Real leadership is simple: take care of your people, enforce the standard, and never ask anyone to do something you won't do yourself.

Courses of Action:

- Perform like your people are watching - because they always are.
- Take care of your people. If they fail, it's a DIRECT reflection on you. Own this.
- Don't avoid conflict - fix it directly. Weak leaders hide.
- When you're unsure, prioritize mission, team, and integrity. Everything else is noise.

Adapt and Overcome

BLUF: Conditions will change without warning - the ones who last are those who keep calm, adjust course, and keep moving the mission forward.

What Are We Missing?

At my first diving unit, one of the Coast Guard's first full-time diving units, Regional Dive Locker East in Portsmouth, VA, I had a salty CWO who had been a Coast Guard Diver for a long time - like diving in the 1980's long. [Fun fact: The Coast Guard has maintained a Diving element since the 1940's, starting as the first frogmen of the OSS.] He was very experienced in Coast Guard operations, having worked over an almost 30 year career on all types of Coast Guard Cutters and units, and for most of that time he was a Diver on those ships as well. He had more experience underwater than most of us had in the Coast Guard. During every planning session, he would always end with, "What are we missing?" This train of thought is so beautiful it has stuck with me to this day. It has, without a doubt, saved me on more than one occasion, as well. It provides an open mind to hear from your team or others on things that you might not have thought of, or things that you might not be seeing. Most decisions in our line of work should not be made in a vacuum. You should seek counsel from other experienced peers, leadership, or teammates, especially if the safety of your team or others is at risk.

For example, when you're planning a complex hoist from a helicopter to a vessel in a high sea state, you need every perspective. The pilot sees it from above, the flight mechanic sees it from the cabin door, the rescue swimmer sees it from the hoist cable, and the coxswain on the boat sees it from the helm. Each person has a unique view of the scene, and your plan needs to account for every single one of those perspectives. By asking a simple, open-ended question, you create a space for everyone to contribute their piece of the puzzle. You also open up the floor to hear from people who otherwise may not speak up. You might present your master plan to the team and it may seem like you have a rock-solid plan with all the details worked out. People may be fearful to speak up so as to not look like the sniper in the back of the room trying to make you look stupid, and they might be trying to do you a favor by not speaking up. There are plenty of reasons people don't speak up - and one of them might be you. Maybe you don't actually want input, because this is your master plan. After all, you're the most qualified, most experienced, smartest, and obviously the best-looking person in the room. Why would you need anyone else's ideas when you've already solved everything in your head? That's your ego talking. And guess what -people see it. They keep their mouths shut because they know your delicate little ego can't handle being challenged. On the other hand, by just asking simple questions, 'What are we missing?' 'What aren't we seeing?' 'What are we forgetting?', changes everything. You're giving people permission to think outside the box, poke holes in the plan, and talk about it openly. The answers don't have to be groundbreaking - most of the time you'll get nothing. But every now and then, someone gives you that one thing that makes the plan safer, smarter, or less risky. For a coxswain or a boarding officer, a single piece of overlooked information - a change in weather, a last-minute change in tasking, or a critical piece of un-briefed intelligence on a vessel's occupants - can be the difference between a successful mission and a catastrophic failure. This isn't a game - it's a matter of life and death, and your team deserves to be part of a plan that has been vetted from every possible angle. Plans built on one perspective fail; plans built on every perspective save lives.

Maintenance

The saying goes, "Take care of your gear and it will take care of you." Maintenance is a massive part of taking care of the gear. This could be a weapon system, an O2 system on a helicopter, a P6 dewatering pump, or diver's life support systems - it all requires maintenance that keeps it working in perfect condition. When you need it, you need it to work with 100% certainty and that requires periodic maintenance or

PMS. PMS means Periodic Maintenance System, which is part of the 3-M (Material Maintenance Management) program developed by the US Navy in the 1960's and introduced into the Coast Guard in the 1970's–80's. It's an entire program that encompasses every part of a ship, cutter, helicopter, or system. While the Coast Guard moved away from the Navy standard of 3-M with its own systems, programs, and documents - the roots remain on some systems and the spirit of the program is there. It's operational insurance that your gear or system will work when you need it. PMS is labor intensive, requires a keen attention to detail, and experience with the system you're maintaining. Your maintenance can mean the difference between life and death for you, your crew, or others - it's nothing to be taken lightly, no matter how mundane it might seem.

The skilled maintenance technician is the cornerstone of operational readiness. They are the trusted professional responsible for ensuring that every piece of equipment is maintained to the highest standards, so it works flawlessly when lives are on the line. This is a tremendous responsibility. It's not just about turning wrenches; it's about managing a maintenance program, tracking hours, ordering parts, and most importantly, ensuring every piece of equipment is safe to operate. Unfortunately, with many mundane tasks that you do over and over, people will find ways to take shortcuts. These shortcuts are lazy, can be deadly, and in the end are not worth the extra time you might gain. A common term you'll hear for poor maintenance practices is "gun-decking" -

claiming a task was completed when it wasn't. The phrase traces back to naval lore: officers who didn't want to climb to the main deck to take a navigation fix, whether because of bad weather, laziness, or being too busy, would instead stay below on the gun deck and enter a false fix in the ship's log. This became known as "gun-decking" the log and has gotten many military personnel in hot water over the years. You don't want to become a part of this club. It's lazy and dangerous. When assigned your duties, especially maintenance, you absolutely need to take the time to complete these tasks step-by-step, with every tool it says to use (including calibrated specialty tools when needed), without skipping steps or jumping ahead, in order to complete it. Does it take time to gather all the PPE, pull the correct tools, tag out the equipment so it can't operate, get the proper cleaning solutions, greases, and lubricants, break it down, do the work, replace the parts, reassemble it, and then untag it and log everything - YES! That's why people cut corners, skip steps, and gun-deck the job - it's tedious. But the tedious part is exactly what makes it work. Don't be a dirtbag - do it right.

Reality Check: If I didn't drive the point home - other people are entrusting you with their safety and you should NOT take this lightly. You can, and plenty of people have, killed themselves, their crew and others by not conducting proper maintenance. You should have <u>zero tolerance</u> for anyone who cuts corners on maintenance. The standard is all or nothing - either it's done right, or it's wrong. If you catch someone on your team gun-decking maintenance, you need to hammer them and hold them accountable - grace doesn't apply here. Lazy people can get good people killed. Not on your watch.

The Snowball Effect

When things go wrong on an operation, it's rarely one big thing. It's almost always a combination of small failures that build on each other until they become a major incident. From experience, there are always signs, and if you're not paying attention, you'll miss them. You have to be aware and use your gut to notice when things are getting sideways. I've often felt like we were swimming upstream just to keep things on track. Everything would start to break, go wrong, or just feel off. That feeling is the gut check - the sinking feeling of impending doom that tells you bad things are coming. For me, recognizing those signs has kept my teams safe in countless dangerous operations.

Sometimes the team just isn't in sync. Most of what we do is a team sport. The stakes are higher than a game, but the dynamic is the same as a professional sports team. You've seen those days when things just aren't "clicking" - one player goes left when they should have gone right, a teammate misses a shot when another is wide open, or someone just isn't playing like themselves. In sports, a bad day means a loss.

In our line of work, a bad day can mean lives are lost. Just like in a locker room, you need someone to step up and give the halftime speech to tell everyone to get their heads out of their asses and in the game. If that's your role - do it. Don't wait for things to go sideways to realize you should have spoken up.

You have to recognize the signs as they appear:

- Pre-op maintenance isn't going smoothly.
- One thing after another is breaking or malfunctioning.
- The weather is changing faster than briefed.
- The team is operating as individuals, not as a cohesive unit.
- People aren't communicating when they should be.
- The on-scene situation is not what was briefed.

The list goes on. It's the leader's job to stop the slide, rally the team, and make the call on whether to continue. There are very few missions so critical that you can justify the gains outweighing the crew's safety. The reality is, those missions are rare and almost nonexistent. Yes, the mission matters and you can never mitigate all risks - but it's our duty to bring our people home safe to fight another day, not push forward when all the signs are screaming that things are about to go wrong.

Hindsight is always 20/20. Post-mishap investigations almost always reveal that all the signs were there. Little things piled up until the snowball became a boulder. You have to be on watch for these signs all the time. Does that mean canceling the operation or training when you see these warning signs? No. Our job is to mitigate risks, as you can never eliminate all risks. If things are breaking before the op, delegate a team to fix them. Check on things, and then make a decision if the fix is safe enough to continue. If the team isn't there mentally, use your voice to gather them up and course correct - get them there.

In military diving, we use a command to get everyone's attention and pass critical information. The diving supervisor will yell, "ON THE SIDE" - and everyone stops talking, pays attention, and waits for the incoming information. For example, the supervisor would say, "ON THE SIDE - Divers moving to the water's edge." Everyone on the "dive side" then responds in unison: "Divers moving to the water's edge." Now everyone is on the same page and knows to be alert and what's happening next. This happens throughout the entire dive from the brief to the very end of the evolution when the call is made, "ON THE SIDE - Divers are clean" (shows no signs or symptoms of an embolism).

In any operation or training, it can be as simple as yelling, "LISTEN UP!" I promise you - heads will turn and you will have the floor. "LISTEN UP - the weather is taking a turn for the worse, tighten up those fenders and be prepared on those lines. Keep your hands and fingers clear and be prepared to slack them when I say." Ask for acknowledgment, "ROGER THAT?" or "COPY THAT?" - get a verbal response. The opposite of this is sitting back, watching the weather get worse, thinking those things to yourself but not saying them out loud, and hoping others notice too. Yet when someone gets hurt or equipment is damaged, it's easy to see plain as day that all the signs were there, and you should have spoken up.

Don't wait until it's too late and the snowball is a massive boulder. Take control, be prepared for things to go wrong, and mitigate the risks.

Decision Making

The Coast Guard is a smaller organization and that's no secret. As a result, the leadership is often handed down at more junior levels. You will, inevitably, be forced to make tough calls on the spot without all of the information and in a short amount of time. Sometimes you'll have to make decisions in the middle of chaos without all the information. How do you make the best call under pressure, knowing the picture is incomplete? The truth is you will almost never have 100% of the information. And if you ever do, it's probably too late - you've already lost the initiative and missed the moment to decide. I was taught by my former boss, mentor, and friend, Master Chief Master Diver Kent "Roc" Robarts USN, (Ret.), that 80% of the information is enough to make a decision and take action. He called it an 80% solution. If there is a problem and we can get it to an 80% solution first, that's all we need to start moving in the right direction and the rest of the information will show itself/work itself out during the process. I watched this happen time and time again with great success. You can easily get bogged down in the weeds by trying to plan for every contingency, every failure and every detail you'll never uncover in time to act. This is called 'Paralysis by Analysis' or 'Analysis Paralysis' and many decisions have been stalled to complete indecision while trying to gather every single fact. Sometimes you have to take a step back, detach from the situation, look at the big picture and recognize what you're ultimately trying to accomplish - and make the most informed decision with the information that you have and keep moving. If you're wrong, cool - change course and keep it moving towards the objective. To further solidify this reasoning of the 80% solution, Jeff Bezos of Amazon fame has said that he only needs 70% percent of the information to make a decision on something (Master Diver was onto something), and clearly Jeff Bezos has experience with moving in the right direction.

Type 1 vs Type 2 Decisions

To expand on his billion dollar decision making process, Jeff Bezos described his decision making framework as Type 1 and Type 2 decisions. I thought this was a good way to look at things so I'll share with you here:

Type 1 decisions are 1-way doors. They are hard to reverse, rare situations that you won't face often and they must be made with great care, deliberation and consultation. These are the make or break decisions and should be made with great care, thought, and due diligence. These are the decisions you DON'T want to make in a vacuum. You need to run these up in the chain of command and get counsel on. There is a saying for these types of things: never be the junior person with a secret! You may think you're doing the right thing by taking on a situation alone - shouldering the weight of something extremely delicate or important - but in reality, you'll likely do more harm than good. Commanding Officers are there to command, make critical decisions, and have accepted command knowing that they will have to bear responsibility for some very hard decisions - give them a chance to make these choices and use them for guidance in these situations, if at all possible.

For a junior member, examples of these types of decisions might be declaring an asset or piece of equipment down (Charlie status) and unusable for a mission, or a boarding officer terminating a voyage and issuing a costly civil penalty. These aren't small choices; they have legal, financial, and operational ramifications. I'm not saying make no decisions or accept no responsibility without running things up the chain, but we all have a part to play. With experience, you will hopefully know and have time to recognize these situations when they come up, and seek appropriate guidance and the appropriate time. CO's/XO's/

Command Staff have lines of communications to people and divisions that we may not have easy access to, such as legal, administrative, or senior command staff. By taking on these irreversible decisions alone, you are not only accepting responsibility for the decision you make but for the outcome and consequences as well. Don't be the junior person with a secret.

Type 2 decisions are 2-way doors that can be easily reversed and these kinds of decisions should be made with relative ease using sound judgment. These are decisions that you should make on a routine basis and make up the majority of decisions you will have to make. You should make these decisions quickly using sound judgment gained from training and experience, as these decisions can be quickly reversed without much harm. If you create a watch schedule for your crew and someone gets sick and you have to switch things around - make it happen and notify your superior after the fact (if you've been delegated that authority). For example, if your Chief doesn't agree with that schedule change, things can easily be moved around - it's not the end of the world.

Reality Check: If you see a Type 1 Decision coming - you need to STOP and seek guidance from the correct person in your chain of command. Why do I say "correct" person? Just because a person has a higher rank, it doesn't mean they have the <u>authority</u> you're seeking. For Type 1, irreversible decisions, you need to "get the monkey off your back" - and seek appropriate counsel, or you will be held personally liable for the consequences. Seeking guidance during a crisis may not be the easy decision, but it very well may be the right one.

Type 1 Decisions: 'One Way Doors'	**Type 2 Decisions:** 'Two Way Doors'
Almost impossible to reverse. You can't go back to the way things were.	Easily reversible. You don't have to live with consequences for long.
Rare	Common
Must be made great care, consultation and due diligence	Should be made quickly by small groups or individuals with good judgment based on training and experience
These decisions carry high stakes and are often irreversible. Aim to get it right the first time because there may be no second chance.	Speed is important - act fast and course correct over time.
Examples: declaring a vessel unseaworthy/terminating voyage (legal ramifications if wrong), failure to launch during SAR (responsible for other's death/injury), signing off on someone's qual when they're not ready (they get everyone hurt, investigation reveals it, you are responsible), gun-decking logs (logs are legal documents).	Examples: changes to watch schedule, delegating tasks to new members, rearranging a workspace, allowing a 'break-in' or U/I (under instruction) to lead an evolution while under supervision, etc.

Resilience

Resilience gets thrown around like a buzzword, but in reality it's about whether you crack when life piles on. It's about being able to perform when you're dealing with a personal crisis, because the mission doesn't stop. The Coast Guard has counseling, chaplains, and peer support. Use them. Acknowledging you have a problem and seeking help is not weakness - it's maturity and strength. It's the only way to get back in the fight. Suicide has claimed more good people than combat ever has, and pretending you're immune is stupidity, not toughness. Your life is worth more than your pride - your team, your family, and this service needs you in the fight. Too often, we amplify problems in our own minds and drive ourselves to catastrophic conclusions. **These storms will pass** if we give them the chance - but only if we give ourselves the time to endure.

Orders will split families, and the service won't care about your marriage or your kid's birthday. The needs of the Service will always come first. That is the contract you signed. You will miss holidays, anniversaries, and graduations. If you don't prepare for that strain, you'll end up with a wrecked home life. A messed-up home life directly impacts your ability to focus on the mission. Communication and planning with your family before orders drop is just as important as packing your seabag. Your family has to be prepared for the realities of service, and you have to do your part to prepare them.

Mental Health Red Flags

Can't sleep: Your mind won't shut down, and your body can't recover. That impacts your judgment and ability to perform under stress.

Rage outbursts: Losing your temper destroys trust. If your crew sees you as unpredictable or unstable, they'll hesitate when you need them most. Discipline isn't just about following rules - it's about showing your people you can be counted on under pressure.

Numbness/apathy: You don't care about the mission or your shipmates. This is a sign of mental exhaustion and burnout, and it makes you a liability.

Heavy drinking: You are using alcohol to self-medicate, which is a dangerous and temporary solution that will lead to more severe problems and an eventual career end.

Resources

Peer support: This is just as important as any official resource. Talk to a trusted shipmate or a Chief who has been there. They will get you the right help, even if it's just to talk. **YOU ARE NOT ALONE.** More people suffer than you realize.

Chaplain (confidential): This is a completely confidential resource, separate from the chain of command, for any problem - not just spiritual issues.

CG SUPRT: A free, confidential, 24/7 hotline with professional counselors available to help with any issue, from family stress to legal or financial problems. Free confidential counseling is available.

Medical officer: A trained professional who can provide a medical diagnosis and help you navigate the system for further treatment.

Family Separation Survival

Schedule regular comms: It is a non-negotiable part of the plan that provides stability for your family.

Set expectations before deployment: Go over the logistics and ground rules with your family, so there are no surprises when you are gone.

Budget for travel, emergencies: Don't let a financial crisis at home become a mission distraction for you.

Keep spouse looped in on orders early: Your spouse or partner needs to be a part of the planning process. They are your teammate in this.

After Action: The ocean doesn't care about your plan, and neither does the enemy. Things will break, weather will shift, and orders will change with no warning. If you freeze, complain, or wait for perfect conditions, you fail. The Coast Guard rewards people who adapt fast, solve problems under pressure, and keep moving when everything around them goes sideways. Flexibility isn't weakness - it's survival. Excuses don't matter, only results.

Courses of Action:

- Expect the unexpected - plan hard, but assume the plan will break the second it meets reality.

- Stay calm under pressure - when chaos hits, breathe, assess, and make the best call with what you've got.

- Keep learning and adjusting - every screw-up, drill, and mission should feed into how you handle the next one.

Learn From the Scars

BLUF: Every mistake is a lesson paid in pain - own it, learn from it, and make sure you never pay for the same one twice.

The 3 B's

During the course of your duties in whatever rate you may be in, whether it's doing administrative work, working in the galley, working in the engine room or in a helicopter, we all have a common bond. We are all members of the United States Coast Guard and that means you are a professional, first and foremost. The uniform you wear carries with it the full weight of the organization's reputation and authority. It's an honor to wear, but it's not a free pass. The public sees the uniform, not the individual, so your actions reflect on all of us.

We are a part of the bigger Coast Guard, the organization, and the broad rules, policies, instructions, and rules of good order and discipline apply to us all. That means everyone - from the most junior seaman apprentice to the most senior admiral. There is no one exempt from the rules, and no one is above accountability. Your job isn't just to do your work; it's to uphold the standards of the entire service.

That requires constant vigilance from us policing ourselves, holding ourselves accountable, and watching out for the pitfalls that can take down our career, or severely stall it by getting in trouble. Nobody is going to do it for you. Your command won't be able to protect you if you make a stupid decision. Ultimately, your career is your responsibility and yours alone.

There are many, many ways to find yourself in hot water while serving in the Coast Guard, as there is plenty of responsibility to be shared. That responsibility is what makes this service so good. We trust our people to perform demanding, high-stakes jobs, often with minimal supervision. You can't have one without the other. The more responsibility you have, the more opportunity there is for something to go wrong - and things do go wrong. You have to be on the lookout for the shortcuts that seem like a good idea at the time but can lead to problems for you in the long run.

I've seen it countless times: someone looking for the easy way out or trying to bypass a process because it's inconvenient. That shortcut, whether it's a quick form, a procedural step, or a simple safety check, is there for a reason - and often written in blood. Ignoring it is a gamble you will eventually lose.

Beware of your worldly desires while in uniform and in a professional organization, because they don't mix well at all. Your uniform is not a costume you can take off and put on a new personality with. It's a professional cloak you wear at all times, even when not on duty. Your actions, on or off base, reflect on the service. This isn't a suggestion; it's a non-negotiable reality of military service.

Watch out for things that are highly valuable to an organization, even though they might not seem like a big deal to you. The Coast Guard is a federal law enforcement agency and a military service. We deal with real money, real lives, real weapons, and real consequences. The things you are entrusted with are not toys. They are assets of the United States government. Treat them as such.

There is a long standing saying in the Coast Guard that you have to watch out for the 3'B's - Boys/Babes, Bullets, and Booze. These three things will get you into hot water faster than anything else, and you are more likely to fall victim to one of these issues than anything else while serving in uniform, based on my personal experience while serving over 20 years (I can't back this up statistically, but please try to prove me wrong, I'd love to hear it). Falling into these issues can derail a promising career, set you back financially by losing rank or punishment of loss of pay, cause you to be discharged from the service, and put an end to your precious time in the World's Greatest Coast Guard.

The service has taken a hard stance on some things, and

these issues (while broadly lumped together) - will wreck your career. The standards are high because the stakes are even higher. I've seen it happen more times than I ever cared to see, and saw some rockstars with promising careers get kicked out, and some even imprisoned. For some of these issues, there is a very tough stance of zero tolerance and if you cross that line - there's no going back. Let me break it down.

Boys/Babes

While serving in the military, you will inevitably work and be around members of the opposite sex. This is no different than any other job or situation that you would encounter in a civilian workplace. You will work alongside together, train together, eat together, travel together, be a part of the same experiences, and get to know your fellow shipmates pretty well in the process. This leads to some people wanting to cross that line of professionalism.

This is a human problem, but in the military, it becomes a career problem. When you are on a small boat crew, a cutter, or a remote air station, your working environment becomes your entire world. The people around you become your family, and you have to be able to trust them with your life. Undermining that trust by trying to start a relationship or worse, a hookup, is a fundamental betrayal of your duty and your shipmates.

You can never allow yourself to cross that line. It is simply not worth it. The second you blur the lines between professional and personal, you have created a liability for yourself, your shipmates, and your command. If something goes wrong, and it almost always does, it will affect your entire unit's morale and effectiveness.

You are not in a speed dating service, or a find-a-significant-other service - you are in the world's most capable maritime service and should not blur any lines when it comes to members of the opposite sex. Do not get disillusioned about your coworkers. Do not get so enamored by your own ego that you think that you are the exception to the rule - you are not. They are there to work, you are there to work - keep it professional at all times and keep it moving.

You will spend long periods working closely with others, going through hard experiences side by side. This usually creates some level of bonding, but you must hold yourself to a high standard. That is the nature of military service - it forges the bonds of brotherhood and sisterhood. Trust and respect are built by relying on each other in difficult situations. Do not confuse that with romantic interest.

You cannot cross the line or allow yourself to think that the line is even crossable. It should not be allowed, tolerated, or considered. Blue on Blue is a no-go. This is a reference to the U.S. military's term for "friendly fire" - when you attack your own. Blue on blue in a personal context means you are doing something that hurts a fellow member of the service, and it's something that will not be tolerated.

While there may be minor exceptions to the rules and fraternization policy such as unit size or separate departments, it is never worth letting yourself play that game. Those exceptions exist for specific reasons and should be respected as written. Do not be the person who tries to push the envelope. The Command is always watching. There are plenty of fish in the sea - go fishing elsewhere. A quick look at this quarter's Good Order and Discipline report says it all - people still get in trouble constantly over this. Most often, it's unwanted advances that bring them down. Let that sink in.

When your ego becomes so large, inflated, and delusional that you convince yourself it's acceptable to cross the line of professionalism, by starting conversations that don't belong in the workplace or by "hitting on" your co-workers - you are being a complete piece of shit. I'm not sugarcoating this. It is harassment, it should never be tolerated, and under the Uniform Code of Military Justice (UCMJ) it is 100% prosecutable. I've seen highly trained, skilled, and knowledgeable members on the receiving end of harassment, and it destroyed their passion and desire to serve in the Coast Guard we all love. This is absolutely unacceptable.

Asking out your fellow service members on dates (regardless of rank/title) should not happen - ever. This includes talking about unprofessional things, eluding to things, making hints, or any other little flirty game you may conjure up in your egotistical needy brain - don't do it. If you keep it 100% professional, you will have no issues here. If you blur the lines, you will have issues, and I will be glad to say I told you so.

Reality Check: This is non-negotiable. I know I'm beating a dead horse here, but for a very real reason - thousands have been punished for crossing this line. No one is immune to hormones, desires, wants, or wishes. It's part of being human, and yes, at the core we all want to be loved. HOWEVER - check yourself. Leave those things at the door when you step onto the deckplate. Exercise ultimate personal discipline and keep work strictly about work. There is no acceptable reason to ever cross this line. It is stupid, it is not worth it, and it will not be tolerated - nor should it be. Consider yourself warned.

BULLETS

I'm not exactly sure who does this now that we have Maritime Enforcement Specialists (ME's) at units now, along with the Gunner's Mates (GM's) - but when I was at my first small boat station, Station Grays Harbor, we had no ME's or GM's to do any of the weapon-type duties. Boatswains Mates (BM's) were responsible for the small arms, pyrotechnics, and ammunition.

That meant as a brand-new non-rate, I was taught by a seasoned BM on how to properly handle, store, and account for every single round. It was a serious responsibility, and even at that low level, it was clear that there was no room for error. The trust placed in us was immense.

When I went through the "striker" program - on-the-job training at your unit that allowed you to complete all requirements to qualify for certain rates without attending an "A" school - a major focus was on weapons, ammunition, and accountability. We followed a Navy system, as is the case with much of the Coast Guard once you peel back the layers. One example was the requirement to conduct a "periodic lot report" (PLR) between the 1st and 8th of every month. That timeline is burned into my memory forever - and if it's changed since, sue me.

This means that every single month you have to account for every single shotgun shell, every single bullet, every single pencil flare, every MK127 parachute flare, and everything that goes bang. The forms were meticulously filled out, signed, and double-checked by a second person. It was a painstaking process, but it drove home the fact that every single piece of ordnance, no matter how small, was a lethal piece of government property.

Again, whether this is still factual or not is beside my point. The lesson behind this count every month was that the service takes these things seriously. It's not about the forms; it's about the mindset. You have to understand that every piece of ammunition, every weapon, and every flare is a tool of lethal force that you are responsible for. 100% accountability for every single round that is fired at the range, every single bullet in inventory, and absolutely zero reasons for anything to go missing and be unaccounted for. I took this very seriously throughout my career and held myself to the same principles that I learned back at the small boat station.

For the first 6 years of my career, I handled weapons on almost daily and I held myself and my boat crew to the same standard that I learned as a young SNBM. When I checked out weapons from the armory at Maritime Safety and Security Team (MSST) 9112 in New Orleans, I used a very important principle that has saved my ass more times than I can count: Trust, but verify. This is an age-old principle that became popular as a slogan said by former President Ronald Reagan in the 1980s. It is perfectly acceptable to trust your fellow shipmates that they are doing the right thing, but it is also fully within your scope of responsibility to verify for 100% accuracy - especially if your name is signed on the dotted line for something. That trust is built on a shared understanding of the seriousness of the task, but the final accountability is on you. If you don't verify, you have no one to blame but yourself.

When you check out a weapon system and the ammunition that goes with it from the armory, both you and the armory personnel sign to confirm the counts are accurate - both for the weapons and the ammo. Now imagine doing this nonstop for weeks during a mission, running 24-hour shift rotations away from your home unit. You're receiving guns and ammunition out of the back of the gunner's mate's truck at the pier, signing for them, and taking them straight down to the boat. Rinse and repeat, day after day, for weeks on end.

Boat crew 1 goes out, then returns to turn gear in or conduct a hot swap of the boat (everyone off, new crew on). Boat crew 2 does the same. And so on, crew after crew. Now it's your turn - boat crew 5 reporting for duty. You're outfitted with an M16 and three 30-round magazines, a shotgun with both slugs and less-lethal rounds, and an M240B with a 7.62mm ammo can already mounted on the bow, plus extra 7.62mm cans stored in the bow deck compartment. Finally, you exchange pistols and pistol magazines with the off-going crew. Now let's play a game.

You head out on patrol, escort a ship safely into port or drop a boarding team onto a vessel, and return to the pier eight hours later. When boat crew 6 takes over, someone decides to 'trust but verify' and checks the M16 magazines - only two are there when there should be three. Uh-oh. You suddenly realize that when you took over the boat from crew 4, you never did a full accountability of the gear, weapons, and ammo. To save time and push forward with the mission, you signed your name on the dotted line confirming everything was there and accounted for - when it wasn't.

Whose fault is it that the 30-round magazine is missing? It doesn't take a genius to see that you are 100% liable and will be held accountable. Can you prove, beyond a shadow of a doubt, that everything was there when you took over? You can't, can you? You can make excuses and point fingers at every other boat crew, but the fact remains - once you sign your name on the dotted line verifying the weapons and ammunition count, the responsibility is yours alone. Do not take this lightly.

The service does not take this lightly. You will be held accountable in some way, shape, or form. It will not be fun, it will not inspire confidence from your leaders, and you will suffer the consequences - whatever they may be. The scenario above is a true story. While in Florida working with the Secret Service during Barack Obama's presidential campaign, our mission was to guard a pier at the hotel and provide a water evacuation point for the soon-to-be President of the United States should the need arise. Not the most exciting assignment - sitting for hours waiting for something to pop off so you can rush a presidential candidate out to sea - but sometimes that is the job. And when it is, you do it to the best of your ability and act as if you are the first line of defense.

When I showed up to relieve another boat crew, I directed my team to conduct a 100% weapons and ammo accountability check - as I always did. That's when we realized a 30-round magazine was missing. Turns out, a crew member who had just gotten off the boat had the magazine stuffed in their backpack. It wasn't on the boat anymore - and it was about to be MY problem had I signed the dotted line. What saved me was following the mantra that has never failed me: "trust, but verify." It saved my six that day, as it has many times since. Being a leader requires trust - but if everything is truly squared away, verifying should never be an issue. Trust, but verify.

Now I've talked a lot about literal "bullets" and ammunition in the previous section - and yes, you should definitely take everything I said seriously, as there is no room for error when it comes to accountability of things that go boom, but the "bullets" in the 3 B's can also mean more. The gear and equipment you are issued, the gear and equipment at the unit, and the supplies that are paid for with funds from the government of the United States of America - are all controlled, owned, and accounted for.

Every piece of equipment is purchased with taxpayer money, and every piece of gear is a part of the mission. Whether it's a shackle for a mooring line or a state-of-the-art GPS unit, it's government property and must be treated with respect and accountability. You cannot simply take, borrow, or steal without repercussion or consequences.

When you are issued personal gear, it's usually documented on a form we "borrowed" from the Air Force - the AF-538 Personal Clothing and Equipment Record. If you're issued Oakley sunglasses or your kit with body armor, it will likely be recorded there. Now, what happens if your unit has an inspection, routine or random, and you don't have the items the government paid for, issued to you, and expects you to maintain? You could be held liable for the replacement cost, held personally responsible for the missing gear, or even investigated by the Coast Guard Investigative Service (CGIS) if the item's cost or type warrants it - night vision goggles, weapon parts, or anything else of high value. Your command and CGIS will decide what's appropriate. I've seen people get burned for this - taking gear they were issued, thinking nobody would notice, and then selling it online.

If you lose your gear that was issued to you, you need to speak up and let someone know or you run the risk of looking like you are concealing the truth. The worst thing you can do is lie about it or try to hide it. That act of dishonesty is often worse than the loss itself and will get you into even more trouble. Things happen in the line of duty, during periods of travel, and while working over the water. I'm not condoning being irresponsible and just losing your gear by any means but shit happens. However, if you lie about it, omit the truth (which is still lying), attempt to conceal the truth (again, lying) - you will be responsible. Be smart.

You may be issued a government travel charge card (GTCC), a credit card issued by a bank with funds linked to a government account, to be used for OFFICIAL travel and while on OFFICIAL orders. This card will allow you to book hotel rooms, eat at restaurants, and rent cars while you are on official duties for the Coast Guard. The credit card is in your name and you are personally responsible for paying the bill.

When you return from travel, you must complete an electronic travel claim and submit receipts for purchases over a set threshold (usually $75 or more, or for specific types of expenses). Once approved, the amount you claimed is deposited into your personal bank account, and it's your responsibility to pay the government travel charge card bill.

Sometimes you'll have the option to pay the card off automatically as the reimbursement hits your account, but I've seen it handled both ways. The system is simple: you travel, you spend money, the government reimburses you, and you pay the government travel charge card bill. Seems like nobody could possibly abuse that, right? Wrong. Since the inception of the GTCC, I've seen people get burned for abusing it - making unauthorized purchases while on travel, using it when NOT on official travel to pay their truck note or rent (both true stories), or flat-out failing to pay the bill after getting reimbursed.

A quick look at the current Coast Guard Good Order and Discipline report shows that the irresponsible government travel charge card users are alive and well. Here are 2 back-to-back examples and I promise I didn't search for more than 2 minutes and found this:

> *i. An E-4 used their Government Travel Charge Card at a strip club for $1798 and the following day, falsely reported the charges as fraudulent. Awarded 30 days of extra duty, a forfeiture of $2569, and a reduction in paygrade to E-3.*

> *ii. An E-5 made 95 unauthorized purchases on their GTCC totaling $2,544.42. Member failed to pay off the balance on the GTCC when instructed to do so. Awarded reduction in paygrade to E-4.*

These offenses happen far more often than they should, and they are an easy way to derail your Coast Guard career. Being a professional Coast Guardsman means being financially responsible. The service will not tolerate missed payments, failure to live up to the core values of Honor, Respect, and Devotion to Duty, or making the Coast Guard look bad in the community by acting like a certified dirtbag. Treat your government charge card as if it were gold-plated and issued by the President of the United States himself. Your command has full visibility into every purchase, can audit you at any time, and you have zero expectation of privacy when using that card.

Travel claim money exists for one purpose - paying off the charge card bill from official travel. It is not a payday, and it is not time to "ball out" when you see extra money in your account. That money is not yours - it belongs to the government - and the government will always get it back. If they owe you, they'll be slow to pay. If you owe them, they'll be lightning-fast to collect.

Reality Check: Don't be stupid - don't use the card for anything unofficial, and don't ever skip paying the bill.

Financial irresponsibility can cost you your security clearance, your rank, or even your career. Letting the bill lapse will also hit your credit score hard, because the account is in your name. These consequences will hang over you for a long time if you play this game - and you will not win. Use the card wisely - only for official government travel - and pay the damn bill in full and on time.

Booze

This last topic of the 3 B's may be the most obvious one - but it gets the most people in trouble by far. When parsing through the Good Order and Discipline data that's provided on a quarterly basis by the Coast Guard Judge Advocate General (JAG), it's clear that alcohol has ruined careers, and continues to do so. Every single quarter, almost without fail, Coast Guard members lose their careers due to alcohol related incidents.

Over the last 5 quarters (at the time of this writing), at least 22 shipmates have had their careers ended or their rank stripped from them, due to alcohol related incidents. That's more than 1 person a month - gone from the service. Every single case starts the exact same way - one drink too many, one bad judgment call and an epic career that takes many years to cultivate is gone overnight. Those people wake up to find out that they will never again wear the cloth of our nation. The uniform that is earned, not given. Everything down the drain. 13 separations and retirements in just over a year - with members and families losing a paycheck, often with no backup plans in place, and a hard dose of reality smacking them in the face. These events are hard to bounce back from and some never fully do.

I'm sure every one of us can name someone whose life has been damaged or destroyed by alcohol in one form or another, and like you, I have my own story. For me, it was Jason "Bear Mountain" Bjornberg - one of the most talented, physically fit, outgoing, likeable, and truly outstanding Coast Guardsmen I have ever had the privilege to serve alongside. He is no longer with us because of alcohol - its effects, its poison, and the irreversible damage it leaves behind. I'm ashamed to admit that I was there, never saw the signs and was definitely in on the action - and wrote it off as "just having fun" or "being Hooyah" (as Divers so often say when they're doing something borderline heroic or borderline stupid). The reality is he had a real and serious problem. He rose to success from being an Electronics Technician (ET) to becoming an MSRT operator back before it was officially called MSRT, when it was still known as E-MSST. He went on to become a Coast Guard First Class Diver (1C), diving supervisor, and deployable team leader, and above all he was the go-to guy to get things done. When the Maritime Enforcement Specialist (ME) rating was created, he transitioned into it because none of us ever imagined the Diver rating (DV) would ever exist.

Later, he went through a divorce, a PCS away from his

kids to TACLET South so he could transition to operational work as an ME, and a downward slide into alcoholism that ultimately ended his life. This was not just another shipmate lost. This was Jason "Bear Mountain" Bjornberg. A man who could outwork, out-think, and outperform nearly anyone, whose presence made every team sharper, and whose absence left a hole that has never been filled. He is still deeply missed by his friends, his family, and those of us who saw what he was capable of and what was taken from him. His death was one of those moments that makes you question everything you thought you knew about life, about strength, and about what it means to be unbreakable. Some people can have one drink and stop. Others are the type that if they have one, it will turn into twelve. If you know you are that type of person, you need to make the decision right now to cut yourself off permanently, because the next drink could be the one that ends you.

I've seen those people ruin everything for themselves and others. If you know you are this type of "all or nothing" person - you need to choose nothing. If you're one of the ones who think they can tolerate alcohol and choose to partake, there is simply no excuse to be irresponsible with it. If you drink - don't drive. Use an Uber or Lyft and don't be an idiot - it is not worth risking your life, your shipmate's lives, or the lives of others.

I watched one of the best Boatswain's Mates I'd ever known ruin his stellar career and get kicked out for alcohol incidents. This guy reached the pinnacle of seamanship in the worst weather and waves anyone would ever care to be in. He was a Surfman in the United States Coast Guard. There is no greater level of maritime skill and mastery for navigation, seamanship, judgment, experience that can be attained. This is the peak. When the seas are too rough for "regular" coxswains - Surfmen take the helm and rescue those in peril, in some of the nastiest weather one can imagine. 30' seas with 25' breaking waves, driving rain, at night with zero visibility - they get the job done. This man had to now find a new career after proving himself time and time again that he was the best - and he was. He was the leader of the pack at the unit. But he had a problem. He may have been lucky for a long time, or kept it under wraps long enough, or maybe it escalated later - but the problem reared it ugly head more than once and the organization he loved so dearly and put his life on the line for time and time again told him to pack his bags and never return. You are no longer wanted in this organization. There is no contesting that once it happens. You can not undo the damage once it's done. There are no do-overs.

This is not a rank-independent issue. The truth is it happens at every level. However, one constant is that the consequences and outcomes are the same, regardless of rank. In over half of the alcohol related incidents, the outcome is a separation from service or a forced retirement (if retirement eligible). Over a seven-year span of Good Order and Discipline reports, alcohol emerges not as an occasional misstep but as one of the Coast Guard's most consistent and unforgiving career-killers. The incidents cut across every uniform in the service - from fresh non-rates who thought liberty was a free pass, to seasoned chiefs and officers with decades of credibility behind them.

In one case, a Chief Petty Officer with more than 19 years in blew a .256 BAC during a traffic stop, failed to notify the chain of command, and was retired in lieu of administrative discharge - losing the chance to finish on their own terms. In another, a junior officer was caught operating a government vehicle after "multiple alcoholic beverages," resulting in relief for cause and the permanent end of their upward mobility. An E-3, just months into their first enlistment, provided alcohol to underage shipmates during a barracks party; the next morning, one of those shipmates was hospitalized for acute intoxication, and the provider was separated from the service before they had even earned a rate.

Other reports show warrant officers removed from duty after a DUI on base, master chiefs relieved for cause after repeat "alcohol incidents," and officers stricken from promotion lists following off-duty misconduct under the influence. There are also the refusals - members who declined breath or field sobriety tests, knowing what the results would be - separation process started immediately as per Coast Guard policy. In many cases, the incident was enough to destroy a career in a single day, regardless of prior record.

> *The Coast Guard requires commands to start the separation process for:*
>
> - *members who get an AI for driving under the influence (DUI) (Ch 1.N.4.b (13) and Ch 2.Q.2.c (2) of CI1000.4C),*
> - *members who get a second AI (Ch 3.G.2 of CI1000.10B), and*
> - *members who refuse a blood alcohol concentration breath test ((Ch 1.N.4.b (13)(b) and Ch 2.Q.2.c (2)(b) of CI1000.4C).*

The consequences are not polite or negotiable: in seven years, at least 59 people were separated or retired early because of an alcohol-related offense. FY2024 alone was a standout year for all the wrong reasons - 56 incidents, 33 of which ended in career-ending separations. That is more than one member every other week losing their livelihood, their benefits, and their standing in the service. And separations are not the only penalty: the reports document reductions in rank that erase years of advancement, reliefs for cause that permanently close the door to leadership billets, and formal refusals to comply with breath tests that carry the same weight as a conviction. These are not abstract statistics - each line is the official record of the moment a Coast Guard career ended, often in a way that left no path back. The common thread is that alcohol moved the decision point. Once that happened, paperwork turned into a permanent record, and the member's future was decided for them. If alcohol makes it into the report, it is already too late - your judgment has been questioned, your reliability is in doubt, and the service will move quickly to protect itself, with or without you.

Reality Check: Alcohol is the fastest way to destroy your Coast Guard career. From FY2018 to FY2024, at least 169 alcohol-related incidents made it into official Good Order and Discipline reports, the kind of cases that do not go away. Fifty-nine members were separated or forced into retirement, and others lost rank or were relieved of duty, ending any shot at advancement. Chiefs, warrant officers, and commissioned officers are right there alongside junior members: DUIs, high BACs, showing up under the influence, providing alcohol to minors, all of it leads to the same place. Everyone always thinks it won't be them. No one in those pages thought it would be them either. Look at the percentage of people who get kicked out over alcohol and one thing is crystal clear: the Coast Guard is not playing around when it comes to alcohol. Don't be stupid and ruin the best job in the world. Be responsible or don't drink at all. This is your warning order.

After Action: Mistakes are guaranteed in this service. Some will sting, some will haunt you, and some will leave permanent scars. What separates the professionals from the quitters is what they do next. Excuses, denial, and cover-ups destroy trust and careers. Owning the hit, fixing it, and carrying the lesson forward is how you earn respect. Every scar is a reminder that you survived, learned, and are harder to break next time. The only failure is refusing to learn.

Courses of Action:

- Own your mistakes immediately - don't wait for someone else to call it out. Take responsibility and move on.
- Debrief yourself and your team - document what went wrong and what will be done differently next time.
- Turn pain into training - share lessons with juniors so they don't repeat your screw-ups.
- Stay resilient - scars mean you've been tested and endured. Wear them as proof you can take a hit and keep going.

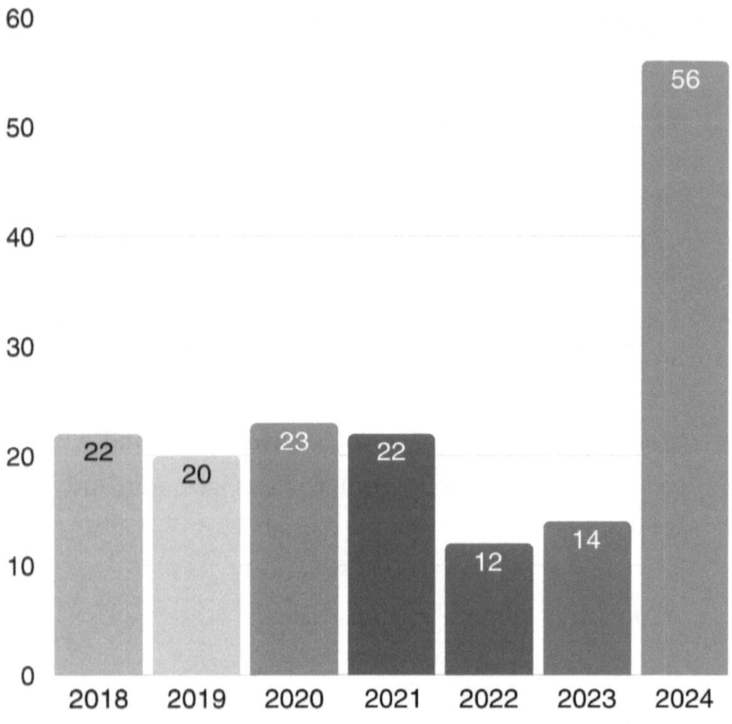

Good Order and Discipline
FY18 - FY24 data

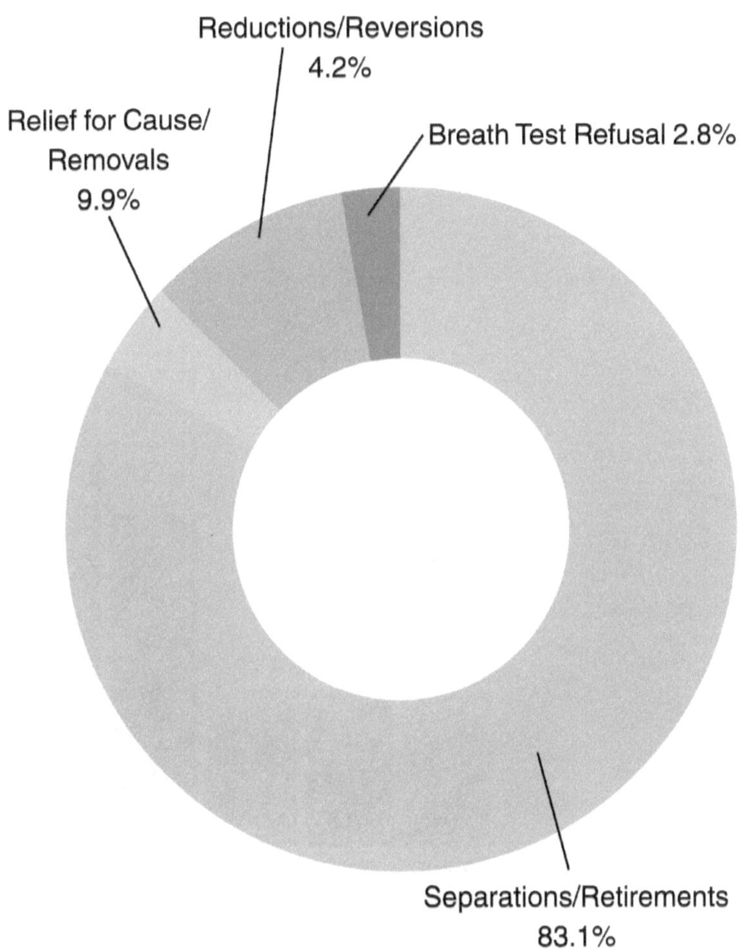

Outcomes of Alcohol Incidents

Good Order and Discipline
FY18 - FY24 data

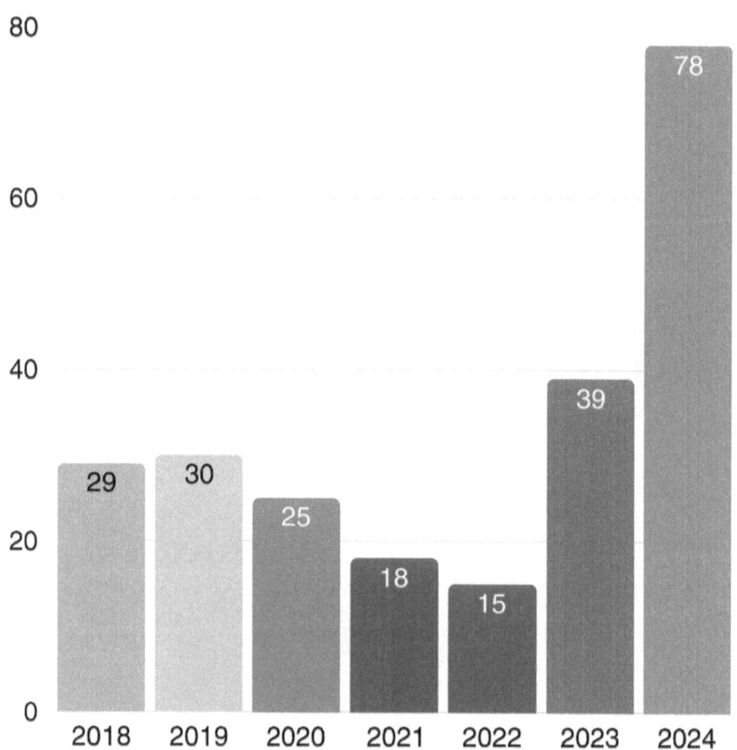

Total Sexual Misconduct and Harassment Incidents

Good Order and Discipline
FY18 - FY24 data

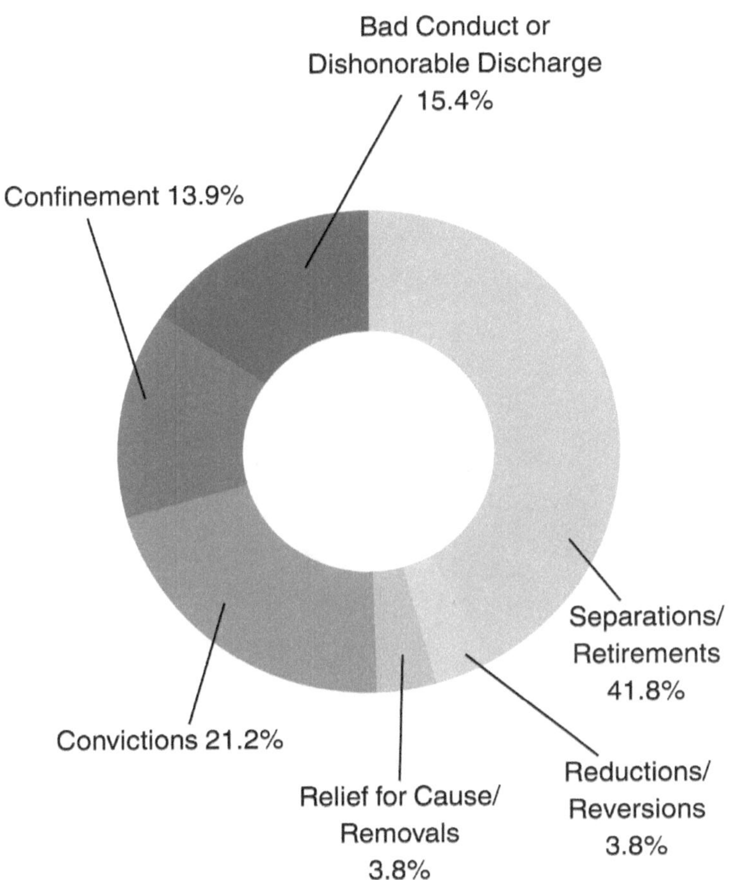

Outcomes of Sexual Misconduct and Harassment Incidents

Good Order and Discipline
FY18 - FY24 data

The Long Game

BLUF: Careers aren't sprints - staying sharp in fitness, finances, and professionalism is what keeps you effective for decades.

Saving For Your Future

You want a future where you're not sweating the small stuff, right? A life where a sudden emergency doesn't send you into a financial spiral. A life where you can actually enjoy your retirement instead of worrying about every dollar. That future starts now, not tomorrow. I've seen too many good people get to the end of their 20 and realize they've got nothing to show for it beyond a pension. That's a mistake you can't afford to make. The TSP - a tool handed to you on a silver platter - is your most powerful weapon against financial mediocrity. It's not complicated. It's not a secret. It's just a damn good deal you need to take seriously from day one. Don't wait until you're a Chief to figure this out; by then you've lost valuable time and money. When do you need to start? YESTERDAY! Put something, anything, into the C, S, or I funds. Set it and forget it. I'm not asking you to become a financial wizard. I'm asking you to make one simple, disciplined choice that will pay dividends for the rest of your life. The mission ahead of you is long, and you need a solid foundation to build on. This is it.

I've seen it firsthand, and I'm telling you - the single biggest regret I hear from retired Chiefs is not starting their TSP early enough. I've seen people retire with over a million dollars in investments - and also seen people retire with $0 - which one do you want? I'm not talking about maxing it out - I'm talking about simply starting. Let's do the math and keep it simple. If you start at 22, put in just $300 a month, and get a modest 8% annual return, you're looking at over $200,000 when you hit your 20-year mark at 42. And that's before a single promotion. If you continue that for another 20 years after you retire, without adding another dime, that same money could grow to over $1,000,000. That's the power of compounding interest, working for you while you're standing watches, deploying, or busting your ass on a maintenance project. Now, imagine if you actually increased your contribution with each promotion. It's not magic - it's just math.

Imagine you're 19, a brand-new E-3. You decide to contribute just 5% of your base pay to the TSP. At first, that's maybe $100 a month. But here's the critical part: you don't touch it, and you let that percentage ride. With every promotion, from E-3 to E-4, E-5, and so on, your base pay increases, which means your TSP contributions automatically increase as well. You're getting raises, and you're putting a piece of every one of them to work for you without even thinking about it.

By the time you're an E-6 or E-7, you're not just contributing that initial $100 anymore; you're putting away a significantly higher amount. Over a 20-year career, with normal promotion cycles, this is not a static calculation. It's a ladder. Your initial small contributions have the most time to grow, and your later, larger contributions supercharge the account. If you average a conservative 8% annual return, which is historically on par with the C fund (S&P 500) over the long term, you're looking at a final account balance well north of $300,000 to $400,000 at retirement.

This isn't a fantasy. It's a proven strategy. I've seen it play out for those who had the discipline to do it. You don't need to be a Wall Street trader. You just need to be consistent and disciplined, the same qualities that will make you a great Chief. The TSP is a tool, just like a wrench or a screwdriver. Learn how to use it now, and by the time you're a senior petty officer, you'll have a retirement fund that gives you options. The stakes are too high to ignore this. Start now, and let time and promotions build your financial security for you.

Now, let's talk about the cost of waiting. Say you're a third-class Petty Officer, and you think you'll "get serious" about your finances when you make Chief. You've already lost a decade. If you wait until you're 32 to start that same $300 a month, with the same 8% return, you'll only have about $80,000 at the end of your 20 years of service. That's a difference of over $120,000. Do you want to give up six figures for no reason? This isn't a training exercise where you get a do-over. This is your one shot to set yourself up. Don't fall for the lie that you can't afford it. You can't afford not to. The pain of putting money away now is a fraction of the pain of having no money in retirement.

The enemy of your financial future isn't the government, your command, or the high cost of living. It's inaction. It's the belief that you have all the time in the world. You don't. The clock is ticking from the moment you hit Boot Camp. Your TSP is a tool of empowerment, a way to take control of a future that is otherwise uncertain. Don't let your service go by and find yourself at your retirement ceremony with nothing but a flag and a handshake. Take this seriously. Start your TSP, put money into it regularly, and let time do the heavy lifting for you. You deserve a solid future, but you have to build it yourself. No one else is going to do it for you.

MENTORSHIP & NETWORKING

No one makes it through a Coast Guard career alone. The Service is too complex, and the path to a successful career has too many hidden turns to navigate it on your own. The sharpest members know how to find mentors - people who've been where you want to go and can help you get there. A mentor is more than just a sounding board; they are a source of hard truths and a guide through the realities of the fleet. A mentor can save you from blind spots, teach you the unwritten rules, and help you avoid mistakes they've already made. The unwritten rules - like how to navigate a difficult command or the best way to get a certain qualification - are learned through experience, and a mentor can pass them on to you. If you're serious about advancement, mentorship is not optional. It is a defining trait of every successful officer and senior enlisted member in the Coast Guard.

You don't need a formal program to find one. The best mentorship is a natural, informal relationship built on trust and mutual respect. Look around your unit. Who's respected? Who gets things done? Who's already in the role you want? The answers to these questions are a direct line to your next mentor. Approach them, ask questions, and show them you're worth investing in. Don't waste their time. Most senior members are willing to help someone who is motivated and humble enough to ask. They see a motivated junior member as a sign of a strong future for the Service and an opportunity to pass on their knowledge.

Networking is just as important. It's the difference between being a good operator and a respected professional. The Coast Guard is small, and your reputation will move faster than you do. The network of Chiefs and Officers is tight. They talk to each other. People talk. When you screen for a special assignment, a competitive school, or a new unit, your future command will call your old commands to ask about your work ethic, your attitude, and your character. If you're known as squared away, opportunities will find you - special assignments, schools, deployments. Your name will be on the short list for the best jobs. If you're known as a dirtbag, doors will close before you even know they existed. You will be non-recommended for assignments you never knew you wanted.

Mentorship and networking aren't about kissing ass. They're about building a professional community and a reputation you can be proud of. They're about surrounding yourself with people who push you to be better and opening doors that would otherwise stay locked. Don't wait until you're a first class or Chief to figure this out. The professional habits you build as a junior member will define your entire career. Start now.

Leaving The Service

Being a part of the United States Coast Guard is a unique and special opportunity. It can take you places that you never imagined going, have you going on adventures that you might only see in the movies, and will have you doing things that you never thought you would ever get to do. The lifestyle comes with adrenaline-filled times, some high-highs and hopefully not too many lows. You'll be challenged in ways you weren't challenged before, which means personal and professional growth for you if you rise to these challenges. Even if you didn't successfully complete every obstacle that came your way, you may not realize it but you will have learned some valuable lessons and insights from being a part of these experiences. While you're in, time seems to fly by - because you're always busy. We have to "do more with less" due to the small size of the service and the number of people in the organization, and that keeps you extremely busy, as it seems as if there is no shortage of things to do. You have to choose your battles and "control your bubble", as I used to say - control the things you have control over - or you will get burned out trying to change the world, so to speak. After all, it is a government organization and the wheels of change move slowly sometimes, as there are definitely politics involved at the top.

In a multi-mission maritime service, you get the chance to get training on a skill set and perform jobs that you may never get the chance to do anywhere else in the world, and that in itself provides some unique challenges. There are not many job opportunities in the world that will pay YOU to attend training, care about you and treat you as a person and not just another employee/number, provide you with top-of-the-line and often state-of-the-art equipment and gear, mentor you throughout your entire career, provide promotion and career opportunities for all, and continually trust you to go out and do these very important and often dangerous things on a regular basis. These opportunities make being in the Coast Guard a highly desirable place to work, even if you can't recognize it at the time. It's easy to take your military career for granted while you're deeply involved in it and so far removed from "civilian" life. It's easy to forget how amazing of an opportunity it is to be a contributing member of our nation's premier maritime service and the World's Greatest Coast Guard, and easy to downplay the epic journey you're on - while you're on it. It's hard to live in the moment and take it all in when the stress of the job has you wound a little tight, when there are a million things to do but you have choose 10, and when you're constantly on the move from state to state, going from mission to another mission, and going from this training to that training. If you're there - take a breather, soak it all in - because that's the beauty of it all. If you want to be there - what are you waiting for? I've said it before and I'll say it again - being in the United States Coast Guard is the coolest job in the world. I never wanted it to end. It's an adventure of a lifetime but the most unfortunate part of the adventure is - it has to end at some point. 20 years sounds like an entire lifetime when you're 18, 19, or 22 years old. Then when you hit the 20 year mark you're like, "where the hell did time go?" Time flies when you're having fun!

Knowing that your journey has to end at some point is actually a positive if you look at it that way. Better to know that your journey has a finite shelf-life than to assume you're going to be around for a while and then you get notified your services are no longer needed. That's what civilian life is like. You show up one day and see an unscheduled meeting on your calendar. They tell you the workforce is being reduced and your services are no longer needed. It's already happened to me once since retiring. Profits are king and people are second in corporate America. Doesn't matter if you're prepared or not - you're taking the ride. In the Coast Guard, unless you screw up and cut your career short, you have a timeline you can plan around. Four-year enlistment - decision. Four-year re-enlistment - decision. Twenty years of service - decision.

If you're a solid performer - doing good work, hitting your wickets (time in rate, time in service, advancement eligibility, etc.), and there's upward mobility in your rating - you're welcome to stay in until you max out your paygrade or become retirement-eligible. The beauty of this is that you can plan your exit when the time is right for you. But the ride does end, and it would be foolish not to prepare for it. Yet time and time again, people reach the end of their enlistment with zero education, zero certifications, and zero plan. If that's how you roll - living as a free spirit who lets the world decide - so be it. It's your life, and free will at its finest. For the rest of us, we prepare. We take fate into our own hands when we have the chance. We live the motto: "No Excuses - Always Ready!" Fail to have a plan, and you've already planned to fail. Take every opportunity to position yourself for your eventual exit from the Coast Guard - not just for your own future, but for the people that depend on you.

The fact is, some of you/us will do highly specialized, military-only jobs while serving in the Coast Guard. Many of these do not translate well "on the outside." You'll have to stretch far and wide to make some of the skills apply. I'm not saying it's impossible, but it's on you - nobody will do the work for you. Go on a job site and type in "Helicopter Rescue Swimmer," "Deep Sea Diver," "Maritime Law Enforcement Specialist," or "Yeoman." You get the picture. These are critical roles in our organization, but they often don't translate directly to civilian careers. That's where preparation comes in. If you think job offers will fall into your lap because you earned five Achievement Medals, an Air Medal from a rescue in the frigid waters of Alaska, became the unit expert on travel claims, or can calculate decompression stops for a high altitude dive op - you're badly mistaken. Those accomplishments matter to us - they are our livelihood - but a civilian recruiter will give you the thousand yard stare the moment you bring them up. They simply don't translate.

You have to decide what you want. If you don't make a conscious choice, you'll end up doing whatever comes your way. Some of you might get lucky and land an opportunity most people dream about, with no worries once you get out. Those people are the ones who step in a pile of dog poop and find a $100 bill. They always seem to fail upward. If this is you, you may not even realize it - but trust me, everyone around you does. We're amazed at your nonchalance and your knack for finding luck in uncertain times. If that's you, I applaud you. For the rest of us, we grind. We work our asses off just to get to status quo, then grind harder to climb to the top, one rung at a time. Our strength is perseverance and determination. We don't accept defeat. We hack away at problems until we break them with sheer will. I'm one of those people. I may not be the smartest in the room, but I will always try to be the hardest working. What I lack in natural aptitude or skill, I make up for by studying, learning, and outworking others. That's the approach I took when it came time to exit the Coast Guard. It was a challenge I faced head on.

You can drag your feet on the way out and let the world decide for you - low-paying jobs, a quick "thank you for your service," and a thousand-yard stare when you mention your time in uniform - or you can take the reins and be the master of your own fate. I chose to take things into my own hands. I'm not here to brag, to claim my way is better, or to say I found the key to success. I won't promise that if you do what I did, you'll get the same results. Not at all. I'm simply here to share what worked for me. I've watched plenty of people leave the service, from E-4 to E-9, each with varying degrees of success. Defining your own success starts with defining what matters to you. What are your values? Do you value money over time? Work-life balance over climbing the corporate ladder? Stability in a career over the excitement of high-stakes contract work? Less pay for the security of a government salary, or the high-risk, high-reward path of commission-based sales? Do you want to be home with your family every night, or do you prefer the travel and new experiences that come with work on the road? I could go on, but there are too many variables - that could fill, and probably has filled, an entire book.

I chose technical competence and a career path with plenty of future opportunities. I grew up on the Westbank of New Orleans, Louisiana, raised by a single mom who did her absolute best considering the circumstances, but opportunity was something I didn't have. The Coast Guard gave me a new lease on life: a steady income (not wealthy by any means, but enough to get by), along with training and educational options that hadn't existed for me before. I made use of those opportunities. It took me over eight years to complete my associate's degree, taking a class or two at a time, but eventually I earned an AA in general studies. Years later, once I knew I would retire at 20 years of service, I went back to school using tuition assistance while on active duty. That stretch was tougher. I had a wife and young children, and I was serving as the School Chief Petty Officer of Aviation Technical Training Center (ATTC) Elizabeth City Underwater Egress School, with all the stress that came with the role. Still, you do what you have to do - No Excuses. Work meant high-risk training during the day, family time after work, putting the kids to bed, then grinding on schoolwork until midnight. Rinse and repeat.

I had always been drawn to computers, and I wanted to build a career in IT after my time in the service, so I chose a Bachelor's degree in Cybersecurity and Information Assurance. The degree plan included several IT certifications that positioned me well for retirement. With a year left before leaving active duty, and uneasy about stepping away from what I called the "comfy blanket of the government," I decided to pursue a Master's in Business Administration (MBA) in IT Management. My goal was to strengthen my options and give myself the best chance of success in the civilian world. For all of this schooling, I used Coast Guard tuition assistance and Pell Grants, and I also leveraged CG COOL (credentialing assistance) to earn a networking certification.

The point here is not my life story. The lesson is simple: the service gives you tools and resources you must use while you can. While I was on active duty, the Coast Guard paid for three degrees because I tapped into what was available. That funding is not guaranteed. At times it barely covered college costs. The same could happen to you, so take advantage while it exists. Let the service pay for your degrees, certifications, or training that will prepare you for the future. This is not about gaming the system, it is about using programs that are there for you. Even if you never work in that field, the credentials matter. CG COOL will pay for certifications, but only if you pass. If there is training you want, ask for it. If you don't ask, it's always a no.

Use every opportunity while you wear the uniform. Every bit counts when you transition, and you never know where you will land. Those who ignore these programs often leave with nothing to show for their service and struggle the most. Do not walk away and start from zero. Position yourself to hit the ground running in your next chapter. It doesn't matter if you're building houses, running cables, coding systems, or traveling the world - you need a plan. The same operational mindset you used here applies out there. Use it. Plan your move and own your transition. If you pivot into a new career: take it from me - expect to grind. Programs like Skillbridge can definintely help. I was one of the first in the Coast Guard to *officially* use it, which opened the door to a civilian fellowship and a new career path. [Note: I don't know if it was the letter to my Congressman that did it, but once he inquired, the Coast Guard had to respond - and not long after, the Skillbridge policy was released. Coincidence? I'll never know.] That single opportunity set my post-service career in motion. The path was not easy and the learning never ends, but that is the reality - grind, adapt, and keep moving forward.

There is life after the Coast Guard - and while it may not be as exciting as what we used to do - it comes with both pros and cons. Remember that everything we did in the Coast Guard was built on planning, predicting outcomes, constant oversight, risk assessment, and correcting course when we strayed. Why would we not hold our own lives and goals to the same standard, forgetting everything we were taught? Use the tools in your toolbox. Apply the lessons we learned - some the hard way - to your life while you are in, during your transition, and after you leave the service. Never accept defeat. Never surrender. Never back down. No Excuses, Always Ready!

After Action: A successful career comes from playing the long game. Every year you stay fit, manage your money, and protect your reputation, you stack wins that pay off down the road. The ones who finish strong are the ones who stay disciplined, keep improving, and treat every tour as a chance to build something that lasts.

Courses of Action:

- Make fitness part of your identity. Don't train just to pass a PT test - train so you can lead from the front and stay mission-ready for your entire career.

- Take control of your finances. Build wealth, avoid debt traps, and set yourself up so money never limits your options or your family's future.

- Keep stacking quals, schools, and education. Career and professional development stagnation is self-inflicted.

- Guard your reputation like it's your most valuable asset - because it is. Build trust, deliver results, and let your name carry weight wherever you go.

Conclusion: A Legacy Worth Protecting

BLUF: Being Always Ready means living with discipline, professionalism, and toughness every day so your team and the public can count on you.

Twenty years. That is a long time to do anything, especially in a service as demanding as the Coast Guard. I've seen it all, from the raw recruits who show up unprepared to the seasoned veterans who finish strong. I have had the privilege of serving alongside some of the finest men and women this country has to offer. I have also had the misfortune of watching some of those same people throw away everything they worked for, for a moment of weakness or a failure of judgment.

This book - my book - isn't a warm-and-fuzzy memoir. It is a gut-check. It is the wisdom I wish I had on day one. I have given you the unvarnished truth about what it takes to survive and succeed here, because the stakes are too high for anything less. The ocean doesn't care about your feelings, your rank, or your excuses. It will swallow you whole if you are not ready. The mission doesn't care about your ego. It requires your best - every single day.

I have seen the relief of a father rescued from a flooding boat in the middle of the Pacific ocean. I've seen the grief on the crew's faces when their 18-year-old deck hand went overboard never to be seen again. I was there before Hurricane Katrina hit - evacuating with my unit and my 11-month-old daughter - and I was there after, when it ravaged my hometown. I saw my mother's home destroyed under 12 feet of water, and I went door to door in a town that looked and smelled like a bomb had gone off. I have felt the weight of responsibility for a multi-million dollar asset and the lives of my crew. I have witnessed firsthand how complacency kills, and how a lack of situational awareness can lead to a tragedy that haunts you for life. I have seen the best of what we can be, and the worst. My time in the service took me on a journey from conducting search and rescue in Washington with some of the biggest waves I'll ever see in my life, to being sworn in as a Canadian peace officer on a joint-border law enforcement mission, to becoming Coast Guard Diver #82 and diving at the bottom of oceans and coastal waterways (not the Hollywood stuff - very often in pitch black, zero visibility), to teaching our DOD partner forces Ice Diving techniques underneath frozen lakes in the mountains of Canada and much more. I have done everything from drug and contraband busts on the water, to securing Guantanamo Bay working with JTF GTMO at the height of the Global War on Terror, to working with our partners in the Indo-Pacific, to qualifying as a Mixed-Gas

Diving Supervisor at Naval Diving and Salvage Training Center. I don't claim to be tough, but I've experienced what it's like to hit a wall and still find the strength to carry on when you need to. You do it for the people on your left and on your right - who I would rather die than let down.

My message is simple and unwavering: Do the hard things. Train until it hurts, and then train some more. Hold yourself and your shipmates to the highest possible standard, because the American public expects nothing less. Don't be the person who blames the weather, their shoes, or their lack of sleep for their failures - be the person who owns their mistakes and drives their own success. This service is a meritocracy, and your career will advance based on your performance, not on seniority.

When the uniform comes off, what's left is the reputation you built. It's about what you've done, the people you've helped, and the trust you've earned with your shipmates. Your reputation is your legacy, and it's the only thing that truly lasts. Pride in service only lasts if you actually lived it. It's not something you can fake. You either did the work, stood the watch, and shouldered your share of the load, or you didn't. Your shipmates know the difference, and so does the public we serve.

If you take nothing else, take this: Day 1, show up on time, squared away, ready to listen. That alone separates you from half the crowd. You'd be shocked at how many new members fail at this basic requirement. Showing up 15 minutes early is being on time. Being on time is being late. Being late is unacceptable and an immediate red flag to your command and fellow crew members. Showing up squared away means you've put in the effort to be professional, from your haircut/hairstyle to the shine on your boots. And being ready to listen isn't passive - it's an active, conscious decision to absorb every bit of knowledge and guidance you're given, no matter how insignificant it might seem. Beyond that, the top ten takeaways in this book are your blueprint. Use them. They're not suggestions; they're hard-won lessons from the deckplate. Following them will keep you out of trouble, make you a more effective shipmate, and build the foundation for a successful career.

And when you're ready for more, don't just close the book. Connect - through my site, through mentorship, through the community of people who've walked the path before you. Nobody makes it alone in the Coast Guard. The people you work with - your Chief, your First Class, and your mentors - have seen it all. Don't be too proud to ask for help, for advice, or for a reality check. We continue the legacy through the new members you train, the professionalism you model, and the standards you refuse to lower. You are a link in a long chain of service, and you have a duty to strengthen it.

Protect the legacy of this service by being a professional in everything you do. Take pride in your uniform and your appearance, because it is the first thing people see. Be a go-getter who is relentless about getting qualified, because a team member who is unqualified is dead weight. Master the fundamentals - physical fitness, communication, and decision making - because these are the bedrock of everything we do. Most importantly, protect your reputation and your career by avoiding the pitfalls - because they will end a promising career faster than anything else.

After Action: You chose the Coast Guard, and that means you chose a harder road. Don't waste it. You've got the chance to do work most people will never even understand - saving lives, protecting this country, carrying out missions that matter. That's an honor, but it's also a responsibility. The only way you succeed here is by showing up every day ready to deliver. No shortcuts, no excuses. The Coast Guard demands your best, and if you give it, you'll walk away with a career you can be proud of.

Courses of Action:

- Show up early, be sharp, and stay ready - half the battle is won just by doing the basics right.
- Stay fit and never stop learning - that's how you stay effective when it counts.
- Knock out quals fast and never stop adding skills - the team needs you at full strength.

- Cut out the distractions that ruin careers - keep your eyes on the mission.
- Lean on your mentors, and then turn around and train your replacements - that's how we keep this service strong.
- Live Semper Paratus every single day - it's not just words, it's the standard you promised to uphold.

I leave you with this final, unbreakable truth: "Semper Paratus" is not just a slogan - it is a sacred promise. It is a promise to your shipmates, to your command, and to the American people that you will always be ready. Now it is your turn to live that promise. I wish you nothing but success. Never take the easy way out.

No Excuses, Always Ready. Semper Paratus!

Appendix A:

Quick Reference Field Guide

Physical Prep (60–90 Days Out)

□ Pushups: 50–70 continuous

Pushup Progression Plan

Week	Method	Frequency	Goal
1-2	10 pushups on the hour	5-6 days/week	20 continuous
3-4	15 pushups on the hour	5-6 days/week	35-40 continuous
5-6	20 pushups on the hour	4-5 days/week	50 continuous
7-8	20 pushups on the hour	4 days/week	60-70 continuous

□ Situps: 50–70 in 2 minutes

Situp Progression Plan

Week	Sets	Reps Per Set	Frequency	Notes
1-2	3 sets	15-20	3 days/week	Focus on form and breathing. Shoulder blades to deck, elbows to knees.
3-4	3-4 sets	20-25	3-4 days/week	Increase pace, practice 2-minute tests weekly.
5-6	4 sets	25-30	4 days/week	Train at near-test pace. Add planks for core stability.
7-8	4 sets	30-35	4 days/week	Aim to exceed 70 in 2 minutes. Focus on endurance.

□ Planks - hold for 1 minute, 30 seconds

□ Pullups: 6+ strict deadhang

Pullup Progression Plan

Week	Method	Goal	Frequency	Notes
1-2	Assisted pull-ups (machine or resistance bands)	3 x 8–10	3 days/week	Focus on full range of motion.
3-4	Negative pull-ups (jump to top, lower slowly 5–7 sec)	3 x 5–8	2-3 days/week	Builds strength in lats/biceps.
5-6	Mixed (assisted + negatives)	4 x 5	3 days/week	Start trying unassisted reps each session.
7-8	Full pull-ups	3–5 sets to max	2-3 days/week	Target 6+ unassisted reps.

□ Flutter kicks: 50+ with boots

Flutter Kick Plan

Week	Sets	Reps	Frequency	Notes
1-2	3 sets	20–25 kicks per set	3 days/week	Hands under buttocks, straight legs
3-4	3 sets	30-35 kicks per set	3-4 days/week	Add 10–15 sec "6-inch hold" after last set.
5-6	4 sets	40–45 kicks per set	4 days/week	Progress to boots on for at least 2 sets.
7-8	4 sets	50+ kicks per set	4 days/week	End each session with 30–45 sec "6-inch hold."

□ Run: 1.5 miles in < 11:30

Running Progression Plan

Week	Warmup	Intervals	Notes	Goal
1-2	Jog 5 min @ 4.5–5 mph (50–60%)	4 × 0.25 mi @ 7–8 mph → recover 2 min @ 3–4 mph	Focus on form, pacing, controlled breathing	Build to 1 mile total fast distance
3-4	Jog 5 min	5–6 × 0.25–0.5 mi @ 7–8 mph → recover 2 min	Extend intervals to .5 when comfortable	1.5 miles total fast distance
5-6	Jog 5 min	4–5 × 0.75–1.0 mi @ 7–9 mph → recover 2–3 min	Train near test pace, longer intervals	2 miles total fast distance
7	Jog 5 min	2 × 1.5 mi @ 7–9 mph → recover 3–4 min	Push endurance, simulate test fatigue	3 miles total fast distance
8	Jog 5 min	Full 2-mile run @ 7–9 mph (sustainable pace)	Practice test pace, taper intensity	Run 2 miles continuous at goal pace

Example Weekly Run Structure

Day	Workout
Day 1	Interval session (standard method)
Day 2	Rest or active recovery (light jog/walk)
Day 3	Interval session (slightly longer intervals)
Day 4	Rest OR core/mobility work
Day 5	Interval session (max effort distance)
Day 6	Easy 30–45 min run OR cross-training
Day 7	Full rest

Taper Run Before Test

Timeline	Adjustment
7–10 days out	Reduce run volume by 30–50%
5 days out	Last interval session
3–4 days out	Practice 1.5 mile run (80–90% effort)
1–2 days out	Very light jogging or rest
Test Day	Arrive rested, fueled, avoid new foods/supplements, pace yourself

Swim Progression Plan

Week	Warmup	Main Intervals	Endurance Swim	Notes/Skill Focus
1-2	200 yd easy swim (any stroke, relaxed breathing)	6 × 50 yd @ 70–80% effort, 30s rest	200 yd continuous, easy pace	Focus on breathing rhythm, full exhale underwater
3-4	200 yd easy swim	8 × 50 yd @ 70–85% effort, 30s rest	300 yd continuous, smooth pace	Emphasize body position, reduce drag
5-6	200 yd easy swim	10 × 50 yd OR 4 × 100 yd @ 75–85%, 20–30s rest	400 yd continuous swim	Add planks/ kick drills for core stability; introduce flip turns if pool allows
7	200 yd easy swim	6 × 100 yd @ 80–90% effort, 20–30s rest	450 yd continuous swim	Maintain stroke efficiency at higher pace
8	200 yd easy swim	4 × 100 yd @ goal pace (minimal rest)	500 yd continuous swim at confident pace	Simulate test: calm breathing, efficient stroke, aim for target time

□ Shoulder endurance for IT sessions

Boot Camp Shoulder Prep

Exercise	Sets/Reps/Hold	Notes / Progression
Overhead Hold (DBs/ bar/backpack)	3 × 60–90 sec	Rifle/canteen hold endurance
Front Arm Hold (plate/backpack)	3–4 × 45–90 sec	Anterior delt endurance
Wall Handstand Hold	3–4 × 20–60 sec	Overhead lockout stability
Pike Push-Ups	4 × 8–12	Pressing strength
Push-Up Position Hold	3 × 30–60 sec	Leaning-rest endurance
Farmer Carry (DBs/ jugs/backpack)	3–4 × 30–50 yd	Carry strength & posture
Lateral Raises	3 × 15–20	Shoulder endurance/ burn

Frequency: 2–3 sessions/week (non-consecutive)
Progression: Add 5–10 sec each week to holds, small weight increases every 2 weeks.

Physical Prep

□ Sleep schedule disciplined (2200 lights out, 0500 up)
□ Nutrition cleaned up, no drinking alcohol

□ Hydration in-check

Mental Prep

□ Practice "showing up anyway". Rain, snow, or shine

□ Keep pushing when it hurts

□ It's not about you anymore. Go harder.

First 90 Days at a New Unit

□ Report early, squared away uniform

□ Get PQS/JQR in hand immediately

□ Identify training PO/mentor

□ Shadow qualified members daily

□ Study outside work every night

□ Knock out 10–15 PQS tasks in month one

□ Volunteer for duties, stay engaged

□ Qualified in at least one major role by Day 90

Squared Away Standards

□ Uniform correct, pressed, clean

□ Grooming on point

□ Early for watch, training, muster/quarters, etc.

□ Communicate clearly and professionally

□ Equipment maintained and ready

□ Positive, disciplined attitude daily

□ PT daily

Accountability Drill

□ After every task: Did I do it fully, correctly, safely?

□ If failed: Self-report/take accountability, document, correct

□ After watch: Conduct self-debrief

□ After mistake: Own it publicly, fix it quickly

Team Player Checklist

□ Always early, always prepared

□ Share knowledge, mentor when possible

□ Back up shipmates without being asked

□ Respect all rates and jobs

□ No gossip, no toxicity, no complaining

Leadership Checklist

□ Lead by example, never ask what you won't do

□ Know your people - strengths and stressors

□ Communicate directly and consistently

□ Enforce standards equally

□ Give credit down, take blame up

□ Train your replacements

Adaptability Checklist

□ Expect plans to change last-minute

□ Always have backup plans

□ Mission first even when gear, weather, or leadership fail

□ Stress management tools: PT, sleep, journaling, counseling

□ Manage family stress early and honestly

AFTER ACTION DRILL

□ What happened?

□ What went right?

□ What went wrong?

□ What to do differently?

□ Did I capture/share lessons?

CAREER NAVIGATION CHECKLIST

□ Get qualified early at every unit

□ Maximize evals through performance

□ Study for service-wide exams from year 1

□ Pursue special assignments (recruiting, instructor, joint ops)

□ Network.Your reputation precedes you

□ Plan for civilian transition from the start

□ Stay fit, stay sharp, stay promotable

Disclaimer: I'm not a coach or doctor - this is personal experience only. Use at your own risk.

Top 10 Takeaways

Show up trained.

Get qualified fast.

Hold the standard.

Own the outcome.

Be the teammate others count on.

Lead by example.

Adapt when things break down.

Learn from scars, don't repeat mistakes.

Play the long game.

Protect the Coast Guard's reputation as fiercely as your own.

Appendix B:

Glossary of Terms & Acronyms

After Action – A summary of what happened and what was learned.

A-School – Coast Guard technical training school for specific ratings (jobs). Where non-rates go to become rated.

BLUF (Bottom Line Up Front) – A single-sentence summary that delivers the main point first.

BM (Boatswain's Mate) – A traditional deck rate specializing in seamanship, navigation, small boats, and leadership.

BOSN (Boatswain Warrant Officer) – Warrant officer specialty focused on seamanship, deck operations, and leadership.

Break-in – A member still in training, not yet qualified for a position or watch. Requires supervision.

CC (Company Commander) – Coast Guard's equivalent of a drill instructor at Boot Camp.

Checkride – Final practical evaluation to verify qualification in a position.

Chief – Senior enlisted rank (E-7) responsible for leadership, training, and mentoring of junior personnel.

Courses of Action (COA) – A plan; a set of intended actions to achieve a goal.

Coxswain – The person in charge of operating and navigating a small boat.

DOD (Department of Defense) – Federal agency overseeing the U.S. military.

DOR (Drop on Request) – Voluntarily quitting from a training program.
DSF (Deployable Specialized Forces) – Coast Guard's specialized tactical and expeditionary units.
DV (Diver) – Coast Guard Diver rating; established in 2015. (Author is plankowner of Diver rating.)
Eval (Enlisted Evaluation) – Performance review used for advancement eligibility and career records.
ET (Electronics Technician) – Rate specializing in electronics, communications, and navigation systems.
Hooyah – Motivational shout used in training and operations (borrowed from Navy/SEAL community; used at Navy Dive School NDSTC); used by military divers.
IT (Incentive Training) – Physical training used as correction or reinforcement during Boot Camp or advanced training.
JQR (Job Qualification Requirement) – Unit- or equipment-specific qualification checklist. Similar to PQS but narrower in scope.
JTF GTMO (Joint Task Force Guantanamo) – A joint command stationed at Guantanamo Bay, Cuba.
ME (Maritime Enforcement Specialist) – Rate focused on law enforcement and security operations.
MK (Machinery Technician) – Rate specializing in shipboard engines and mechanical systems.
MSRT (Maritime Security Response Team) – Deployable specialized counterterrorism and security unit.
NDSTC (Naval Diving and Salvage Training Center) – US Navy training center that trains military divers from all

services.

NJPs (Non-Judicial Punishments) – Disciplinary actions handled by command without court-martial.

Non-Rate – An enlisted member who has not yet graduated from A-School (typically E-1 through E-3).

OOD (Officer of the Deck) – Officer in charge of the ship when underway. Equivalent to Watch Officer.

PCS (Permanent Change of Station) – Transfer orders to a new duty station.

PQS (Professional Qualification Standard) – Official qualification guidebook for a position. Contains tasks, knowledge, and performance requirements.

PT (Physical Training) – Exercise sessions, both scheduled and incentive-based.

Quals (Qualifications) – Certifications proving you are trained and trusted to stand a watch or fill a role.

QM (Quartermaster) – Legacy rate specializing in navigation and ship handling. Merged into BM rating.

SAR (Search and Rescue) – One of the Coast Guard's primary missions: saving lives at sea.

Semper Paratus – Latin for "Always Ready," the Coast Guard motto.

Squad Leader – Recruit leadership role in Boot Camp, responsible for managing a small group of shipmates.